Communications in Computer and Information Science 832

More information about this series at http://www.springer.com/series/7899

Alexandra Ioana Cristea · Ig Ibert Bittencourt
Fernanda Lima (Eds.)

Higher Education for All

From Challenges to Novel Technology-Enhanced Solutions

First International Workshop
on Social, Semantic, Adaptive and Gamification Techniques
and Technologies for Distance Learning, HEFA 2017
Maceió, Brazil, March 20–24, 2017
Revised Selected Papers

Springer

Editors
Alexandra Ioana Cristea
FBCS FHEA SMIEEE
Computer Science Durham University
Durham
UK

Fernanda Lima
Ciência da Computação
University of Brasília
Brasilia
Brazil

Ig Ibert Bittencourt
Universidade Federal de Alagoas
Maceió, Alagoas
Brazil

ISSN 1865-0929 ISSN 1865-0937 (electronic)
Communications in Computer and Information Science
ISBN 978-3-319-97933-5 ISBN 978-3-319-97934-2 (eBook)
https://doi.org/10.1007/978-3-319-97934-2

Library of Congress Control Number: 2018950776

This Springer imprint is published by the registered company Springer Nature Switzerland AG
The registered company address is: Gewerbestrasse 11, 6330 Cham, Switzerland

Preface

Economic development depends on highly skilled, educated individuals. Lower education levels are linked to low income and poverty. In the UK, higher education is available to many high school graduates (~50% in 2011–2012 [3]). In Brazil, the numbers are much lower (< 20%; OECD, 2014). Pioneering efforts from academia and the business sector (incl. proposers) to support pupils from impoverished, underprivileged backgrounds to enter universities (to pass the intensive public exam "ENEM") are e-learning solutions. Still, initial data show typical distance-learning problems: isolation and lack of customisation.

This preface presents the main ways these essential problems and issues were tackled via a Newton Research Links Workshop, and subsequently how they were followed up in this book. We explore major themes that emerged from the submissions we received.

The Newton Researcher Links workshop (http://newton.nees.com.br/) invited UK, Brazil, and international researchers to contribute toward solving the urgent, timely problem of access for children from impoverished backgrounds, via new, seamless personalisation, gamification, semantic and social interaction techniques to meet the specific needs of the millions of candidates, based on (big) data analysis, user analytics, scalability. It is dedicated to promoting equal opportunities. This workshop was extremely successful, with excellent discussions and great networking and further funding applications resulting from it. What it showed, however, was that we were only scratching the surface, and that further efforts were needed to address the overall topic of "Higher Education for All," a noble goal in today's society.

Therefore, we proposed this book, *Higher Education for All: From Challenges to Novel Technology-enhanced Solutions*, opening the Call for Papers beyond the original participants, on an international scale. As a result, we received 31 chapter submissions, out of which, with the help of our Program Committee, we were able to select the 12 most relevant and best papers to include in our book (with a 38% acceptance rate).

The selected papers are both broad in scope, representing a good variety of work toward scaffolding and supporting higher education for all, and centred around three major topics that we identified as follows:

- Higher education online around the world
- User modelling and grouping
- Technology-enhanced solutions: gamification and educational games

Thus, the book commences with the overall picture of issues and solutions in higher education around the world, starting with an analysis of pupils' concerns regarding transition into higher education (Chap. 1), then moving on to a cultural view of higher education in the Amazon, including its solutions and challenges (Chap. 2). This topic delves into a more detailed view of the work for higher education, looking into undergraduate students and their language (Chap. 3). Finally, a major area of current expansion in higher education worldwide is tackled by the chapter on helping teachers of massive open online course (MOOC) do their jobs (Chap. 4). Especially in remote areas, MOOC seem to hold the promise of the future, but currently they are not delivering, and teachers (as well as their students) need to be supported.

Among the methodologies for addressing online learning issues in wide settings in particular, personalisation holds the strongest promise. In this context, user modelling (learner modelling) is essential. Thus, the next topic on modelling and grouping users commences with Chap. 5, on group formation in computer-supported collaborative learning (CSCL), dealing with the social aspects of the online learning process. The knowledge acquisition process, another essential process to trace while modelling online users, is dealt with in the next contribution (Chap. 6). This process is made especially interesting when tackling areas of complex nature. Another vital, yet often neglected, aspect of user modelling, especially in an international set-up with learners from many corners of the world, is the cultural aspect. This is targeted in Chap. 7, from the fundamentals to an experiment. Finally, within this topic, the metacognitive ability[1] of learners is also examined: The elusive knowledge of "knowing what one knows" is not only hard to master, but also equally hard to model within an online environment. This challenge is addressed in Chap. 8.

The final, and thus the most modern and relevant topic in terms of technology for online learning in general, and higher education in particular, is that of delivering the knowledge in a fun and interesting way – similar to how a really good tutor would deliver in one-to-one teaching, or in a very small learning group setting. One of the hottest current areas in this respect is related to games: from educational games (i.e., games that have aspects related to education) all the way to gamification (i.e., e-learning systems that may mimic one or more aspects of the game industry, without being games). The motivation behind introducing games, beside the fun aspect, is the hope that, at some point in the future, our e-learning systems will be working in a similar way to computer games, in the sense that learners will be "hooked" to them without any external pressure, and will want to explore and learn more of their own accord, thus being immersed completely in the experience [1]. While there is a great buzz on the topic of educational games and gamification, this topic starts with a word of caution: the dark side of gamification is also explored, in terms of potential negative effects to be avoided (Chap. 9). Next, an actual game used for software engineering education is presented in Chap. 10 to exemplify the area. Following this, the social (and thus, here, collaborative) aspects of the online interaction of students are explored via gamifications, as well as other state-of-the-art technologies, such as ontologies (Chap. 11). Finally, the last chapter in this topic and in the book (Chap. 12) investigates the designer process, to help developers put together such serious game design via agile methodology.

We wish to specifically thank the Programme Committee for their work in making this book a high-quality publication; their priceless comments helped in revising the papers to be both more appropriate and more relevant for the overall topic of this book.

---

[1] http://www.cambridgeinternational.org/images/272307-metacognition.pdf

References

1. Csikszentmihalyi, M.: Flow: the psychology of optimal experience (1st ed.). New York: Harper & Row. ISBN 9780060162535 (1990).
2. OCDE. 2014, Education at a Glance 2014. Technical Report. https://www.oecd.org/brazil/EAG2014-Country-Note-Brazil.pdf, last accessed 2017/11/21.
3. The Guardian: Number of students going on to higher education almost reaches 50%, https://www.theguardian.com/education/2013/apr/24/students-higher-education-almost-50-per-cent, last accessed 2018/03/26.

March 2018

Alexandra Ioana Cristea
Ig Ibert Bittencourt
Fernanda Lima

Organization

Editors

Ig Ibert Bittencourt	Federal University of Alagoas (UFAL), Brazil
Alexandra Ioana Cristea	Computer Science Durham University, UK
Fernanda Lima	University of Brasília (UnB), Brazil

Program Committee

Afaf Alamri	Ministry of Health, Saudi Arabia
Bedour Alshaigy	Oxford Brookes University, UK
Diana Andone	Politehnica University of Timisoara, Romania
Ig Ibert Bittencourt	Federal University of Alagoas, Brazil
Paul Brna	University of Leeds, UK
Hong Bui	University of Bath, UK
Rosa M. Carro	Universidad Autonoma de Madrid, Spain
Daniela Cascini	Federal University of Minas Gerais, Brazil
Thais Castro	Federal University of Amazonas, Brazil
Cristian Cechinel	Federal University of Santa Catarina, Brazil
Stefano Ceri	Politecnico di Milano, Italy
Giser Chalco	University of São Paulo, Brazil
Koula Charitonos	The Open University, UK
Alexandra Ioana Cristea	Computer Science Durham University, UK
Hugh Davis	University of Southampton, UK
Diego Dermeval	Federal University of Alagoas, Brazil
Darina Dicheva	Winston-Salem State University, USA
Fabiano Dorça	Universidade Federal de Uberlandia, Brazil
Jonathan Foss	The University of Warwick, UK
Serge Garlatti	IMT Atlantique, France
Isabela Gasparini	UDESC, Brazil
Fawaz Ghali	Manchester Metropolitan University, UK
Michail Giannakos	Norwegian University of Science and Technology, Norway
George Gkotsis	KCL London, UK
Sabine Graf	Athabasca University, Canada
Suncica Hadzidedic	University Sarajevo School of Science and Technology, Bosnia-Herzegovina
Maurice Hendrix	Coventry University, UK
Eelco Herder	Radboud University, The Netherlands
Sharon Hsiao	Arizona State University, USA
Jessica Humphreys	The University of Warwick, UK
Seiji Isotani	University of São Paulo, Brazil
Ilkka Jormanainen	University of Joensuu, Finland

Mike Joy	The University of Warwick, UK
Judy Kay	The University of Sydney, Australia
Mizue Kayama	Shinshu University, Japan
Emma King	The University of Warwick, UK
Kinshuk	University of North Texas, USA
Anelise Kologeski	Federal University of Nova Hamburgo, Brazil
Sebastien Laborie	T2i, LIUPPA, Université de Pau et des Pays de l'Adour, France
Katherine Maillet	Institut Mines-Télécom, Télécom Ecole de Management, France
Claire Mann	University of Nottingham, UK
Ivana Marenzi	L3S Research Center, Germany
Scheila Martins	University of Coimbra, Portugal
Eleandro Maschio	Universidade Tecnológica Federal do Paraná, Brazil
Riichiro Mizoguchi	Japan Advanced Institute of Science and Technology, Japan
Andreea Molnar	Lancaster University, UK
Bernardo Nunes	PUC-Rio, Brazil
Elaine Oliveira	Federal University of Amazonas, Brazil
Daniel Onah	The University of Warwick, UK
Alvaro Origosa	Universidad Autonoma de Madrid, Spain
Ranilson Paiva	Federal University of Alagoas, Brazil
Hazel Price	University of Cambridge, UK
Tiago Primo	Federal University of Pelotas, Brazil
Alaa Qaffas	King Abdulaziz University, Saudi Arabia
Dana Qudah	University of Jordan, Jordan
Tiago Roberto Kautzmann	Unisinos, Universidade do Vale do Rio dos Sinos, Brazil
Rafaela Rocha	University of São Paulo; Federal University of ABC, Brazil
Joshua Scotton	True Fit Corporation, UK
Lei Shi	University of Liverpool, UK
Júlia Silva	Federal Institute of Rio Grande do Sul, Brazil
Alan Silva	Federal University of Alagoas, Brazil
Jane Sinclair	The University of Warwick, UK
Sibel Somyürek	Gazi University, Turkey
Sergey Sosnovsky	Utrecht University, The Netherlands
Michael Spector	University of North Texas, USA
Natalia Stash	Eindhoven University of Technology, The Netherlands
Karen Stepanyan	London School of Business and Management, UK
Craig Stewart	Coventry University, UK
Carlo Strapparava	FBK-irst, Italy
Elaine Tan	Newcastle University, UK
Tara Tressel	McGill University, Canada
Ari Tuhkala	University of Jyväskylä, Finland
Mark Zarb	Robert Gordon University, UK
Yiwei Zhou	HSBC, UK

Contents

Higher Education Online Around the World

An Analysis of Pupil Concerns Regarding Transition into Higher Education

Mark Zarb[1(✉)] and Angela A. Siegel[2]

[1] School of Computing Science and Digital Media, Robert Gordon University, Aberdeen, UK
m.zarb@rgu.ac.uk
[2] Faculty of Computer Science, Dalhousie University, Halifax, Canada
siegel@dal.ca

Abstract. Transitioning to higher education is often a stressful experience, with incoming students facing similar issues year after year. This chapter presents two years of data collection regarding the concerns of Computing secondary school pupils when considering their upcoming transition into the first year of higher education. Over the two-year period, it can be seen that pupils continue to demonstrate concerns regarding topics related to money, jobs and course achievement as opposed to those related to environment or social issues. The consistency between relative areas of concern over the two years is striking, further suggesting that an understanding of these issues might help higher education institutions to better support their incoming students.

1 Introduction

Typically, the literature deals with issues relating to transition into higher education from the perspective of early undergraduates (Lowe and Cook 2003; Krause and Coates 2008), to better understand the issues and concerns that arise when transitioning from secondary school into a higher education environment. This chapter will also discuss implications on distance learning environments, and how these concerns are relevant to tutors and coordinators for distance learning courses.

2 Background Work

Transitioning into higher education is a very individual experience. Some students view higher education as an unknown entity, especially if they are the first in their family to go to university (Askham 2008). These students are found to lack the cultural capital and support system needed to adequately access teaching and learning within higher education environments (Leese 2010), as incoming students are repeatedly facing the same issues, especially within STEM (Science, Technology, Engineering and Mathematics) subjects (Hulme and De Wilde 2015).

A. I. Cristea et al. (Eds.): HEFA 2017, CCIS 832, pp. 3–16, 2018.
https://doi.org/10.1007/978-3-319-97934-2_1

The issues faced by these students have been discussed in a number of publications. (Yorke 2000) presents instances where students are shown to experience issues associated with financial management: students started their studies blind to this issues, and were only subject to them once they had started their studies. Ozga and Sukhnandan (1998) show that students experience several issues with managing their new lives within their new environment, both in terms of having moved far from home to a city that may be larger and busier (especially for students coming from more rural areas), and also due to the extra independence that they are suddenly awarded.

In the UK, it was reported that in the academic year 2013–14, Computer Science-related subjects suffered from low continuation rates - that is to say, 9.8% of first year students reading for a Computing-related degree failed to progress from their first year of study. This rate was the lowest across all subject areas in the country, and well above the average rate of non-continuation (6%) (HESA 2016). This data is alarming, as it shows that even if as higher education providers, we are able to get students in the door, this has no impact on retention with regards to continuing their studies. Whilst this issue may seem to be an academic one, the authors posit that there could be additional factors at play: research shows that students may drop out due to the perceived difficulty of the course, difficulties with time management and independent learning, and a low comfort level with the 'new' environment (Kinnunen and Malmi 2006).

A previous study by Siegel and Zarb (2016) reports the results of a survey carried out with 249 school pupils in the 2014–15 academic session about their concerns with regards to their upcoming transition into higher education Computing. The study showed that pupils were highly concerned about certain aspects of transition, including financial implications, what jobs would be available upon graduation, and what implications potential academic failure would have upon these aspirations.

The survey, and the reported conclusions, form the basis for this chapter.

3 Study

The primary objective of this research is to identify what concerns were reported by students in the 2015–16 academic session, and to establish what correlations and differences exist between that data, and data collected from a different population in the 2014–15 scholastic year. The analysis and comparison of the two sets of data will allow for an understanding of any emerging trends and ongoing issues to be developed, and recommendations for improvement to be made.

3.1 Institutional Context

The Robert Gordon University (RGU) is a public research university based in the North East of Scotland, with over 17000 students. It is one of the most northern universities in the UK, and attracts a number of students from more rural communities. Within the School of Computing Science and Digital Media, students study a number of modules per semester, with two semesters spread across one academic year.

Within Scottish higher education, home students (at the time of writing, a classification consisting of students from Scotland as well as the European Union) are typically eligible to have their tuition subsidised by the Students Awards Agency Scotland (SAAS), effectively allowing a fee-free degree. Students who wish to study at undergraduate level are typically required to complete a set of national exams colloquially known as 'Highers' to meet a university's entry requirements - these can vary by department and by institution. These exams are taken by most students at age 16–17.

3.2 Participants

Through the use of mailing lists, social media and departmental contacts, a number of teachers across Scotland were contacted, requesting their pupils' participation. Teachers that replied to the first contact were sent a more detailed proposal, outlining the aims and objectives of the study. Teachers whose classes had previously participated in the study were asked to exclude repeat participants so as to avoid bias due to incomplete follow-up. Furthermore, this allowed for the survey to be replicated more widely, allowing for a larger user base to be gathered.

A total of 307 students from 16 secondary schools across Scotland agreed to participate in the data collection exercise, which was carried out over the course of a few months. This was an increase over the number of participants from the previous session. Three of the schools had taken part in the previous session. All pupils were asked to anonymously fill in a survey which aimed to gather student concerns. To allow for comparisons between years, the same survey was used as in the previous session of data collection.

3.3 Method

Whilst in the previous session participation was only available following face-to-face sessions, due to feedback from more rural communities and due to restrictions within the schools' timetables, it was decided to also allow participation through an online survey administered via Google Forms. Teachers were asked to select their preferred method of participation when signing up to the study. In either case, the participant pool was limited to students in their final year of studies pre-higher education, as these were considered to be on the cusp of transition. Both versions of the survey consisted of the same questions.

All teachers opted to use the online sessions for data collection, and in some cases, requested that academics visited the school to provide context.

Face-to-Face Sessions. Academics from RGU arranged to visit each school to deliver a lecture on a topic agreed with the teacher in advance - these topics ranged from project management, to transitions, to a generic 'life at university' talk. Prior to the lecture, the academic asked each pupil to anonymously fill in the online survey which aimed to gather student concerns.

Online Sessions. All invited teachers were sent a link to the Google Form and asked to distribute this among their class during a supervised session. The online sessions were well received as they required less buy-in time for teachers. Furthermore, due to the framework in which Google Forms operate, the researchers were able to easily send each school anonymised automatically-generated charts summarising pupil concerns, thus allowing for teachers to organise feedback sessions with their pupils.

3.4 Survey

The survey delivered was the standard survey used in the previous study (Siegel and Zarb 2016), consisting of Likert-scale questions grouped into the following larger topic areas:

- Academic Environment;
- Academic Staff;
- Academic Work and Workload;
- Accessibility;
- Homesickness;
- Housing;
- Job-related Concerns;
- Money; and
- Social Concerns.

The aim was to gather anonymous responses from high school pupils regarding their concerns when faced with the transition to higher education, specifically within Computing-related subjects. Due to the change in delivery format, a note was appended to the start of the survey instructing pupils that all questions were optional, and that no identifying data would be collected.

All collected data was anonymous - the only identifier was the pupils' school name, as this allowed for a geographical visualisation of the data, and for the author to feed data back to the various schools upon request. It is currently anticipated that trends and patterns could be mapped to certain geographical communities (for example, would concerns differ between rural and urban schools, or schools with differing socioeconomical status?). This topic is intended for future discussion, and outside the remit of this paper.

4 Data Analysis

This section will discuss the gathered data, grouped by question. Within each section, mean (M) and standard deviation (SD) values are reported for each question, across both years of data collection. Within the tables, Year 1 represents data collected in the 2014–15 session, and Year 2 represents data collected in the 2015–16 session. For each concern, M ranges from 0 (No Concern) to 4 (Major Concern).

Whilst data from both years is collected and presented, the purpose of this exercise is not to statistically compare and contrast individual data points across years, but to understand emerging trends and to present what recommendations can be made with regards to the self-reported concerns about the pupils' upcoming transition to higher education. This understanding is presented in Sect. 5.

4.1 Academic Environment

This category consisted of three questions. The mean and standard deviation of each question for both years of collection are given in Table 1 below.

Table 1. Academic environment: M & SD

Concern	Year 1		Year 2	
	M	SD	M	SD
Class size	0.9	0.86	1.4	0.95
City size	0.9	0.87	1.3	0.94
Lecture or lab environment	1.0	0.86	1.6	0.90

It can be seen in Table 1 that there is little concern reported across the board for questions on Academic Environment.

This is consistent with data collected in the previous session, suggesting that although much effort is placed into ensuring effective learning environments at the higher education level, these items are of little concern to school pupils when considering their transition.

4.2 Academic Staff

This category consisted of three questions. The mean and standard deviation of each question for both years of collection are given in Table 2 below.

It can be seen in Table 2 that overall, pupils have reported some concern with regards to academic staff. The authors posit that this concern is largely due to the fact that pupils are considering the unknown: whilst they have had time to familiarise themselves with their current teaching staff, a transition to higher education presents a large change in terms of staffing.

It is interesting to note that pupils are most concerned about the availability and interest of staff, suggesting that pupils may be concerned about how approachable academic staff might be in higher education.

Table 2. Academic staff: M & SD

Concern	Year 1		Year 2	
	M	SD	M	SD
Will the teaching staff be friendly?	1.2	0.87	1.8	0.89
Will staff be available to help?	1.4	0.90	2.0	0.81
Will teaching staff be interesting?	1.4	0.95	2.0	0.79

4.3 Academic Work

This category consisted of seven questions. The mean and standard deviation of each question for both years of collection are given in Table 3.

Table 3. Academic work: M & SD

Concern	Year 1		Year 2	
	M	SD	M	SD
Will I choose the right course?	1.7	0.95	2.2	0.82
What will the workload be?	1.6	0.82	2.1	0.66
Will I manage my time well on my own?	1.5	0.93	2.0	0.79
Will I like the course?	1.5	0.97	2.1	0.82
Will I be good at the course?	1.6	0.89	2.2	0.77
Am I prepared for the course?	1.6	0.87	2.1	0.74
Will I fail, and what happens if I do so?	1.8	0.95	2.2	0.81

It can be seen in Table 3 that pupils rated all issues highly in terms of concern. As with the previous year, it is clear that pupils are worried about failing a course that they have not yet embarked on, and this is concerning from an academic point of view as this causes unnecessary anxiety early on in their university careers.

Other issues, such as the choice of correct course, preparedness, workload and likability are perhaps natural at this stage in the pupils' academic career, and there is little that can be done to mitigate these issues other than to provide clear and accessible information to pupils about their upcoming course.

The concern regarding time management is worth exploring further, as students are expected to be more independent in their learning once they reach higher education, but may not be sufficiently prepared for this increase in responsibility.

4.4 Accessibility

This category consisted of one question. The mean and standard deviation for both years of collection are given in Table 4 below.

Table 4. Accessibility: M & SD

Concern	Year 1		Year 2	
	M	SD	M	SD
Will there be adequate disability support?	1.1	1.35	1.3	1.01

It can be seen in Table 4 that there is little concern regarding Accessibility reported by pupils, as with the previous year.

The authors posit that whilst the numbers indicate low concern, pupils who did not need disability support would have rated this low on their list of concerns. Furthermore, pupils who did need disability support may have had their parent or guardian organise this support with the school, and may not have realised that this would change when accessing higher education.

4.5 Homesickness

This category consisted of four questions. The mean and standard deviation of each question for both years of collection are given in Table 5 below.

Table 5. Homesickness: M & SD

Concern	Year 1		Year 2	
	M	SD	M	SD
Missing my friends/family/pets	1.1	0.92	1.5	1.01
Distance from home	1.0	0.89	1.3	0.99
How often can I go home?	1.2	0.93	1.5	1.01
When can I go home?	1.1	0.88	1.5	1.01

It can be seen in Table 5 that there was little concern reported on the topic of Homesickness, consistent with results from the previous year.

It is interesting to note that some concern is reported for the 'distance from home' question, considering this is an area that pupils have had a certain degree of control over. Whilst there is little that can be done with regards to the actual feeling of being homesick (i.e. missing family and friends), it can usually be fairly straightforward for pupils to find out when they would be able to go home by referring to their institution's academic calendar. The authors posit that if this information is made available to students at application stage, that this might lessen this concern.

4.6 Housing

This category consisted of two questions. The mean and standard deviation of each question for both years of collection are given in Table 6 below.

Table 6. Housing: M & SD

Concern	Year 1		Year 2	
	M	SD	M	SD
Finding good quality of housing	1.5	0.83	2.0	0.87
Living with flatmates	1.4	0.91	1.9	0.93

It can be seen in Table 6 that pupils have reported concerns with regards to Housing.

For a vast majority of pupils, this is likely the first time that they would be finding alternative accommodation, and be living with people outside their immediate family.

4.7 Job-Related Concerns

This category consisted of two questions. The mean and standard deviation of each question for both years of collection are given in Table 7 below.

Table 7. Job-related: M & SD

Concern	Year 1		Year 2	
	M	SD	M	SD
Will I find a part-time job?	1.5	0.94	2.2	0.76
Will I get a good job after graduating?	1.9	0.96	2.4	0.72

It can be seen in Table 7 that pupils have reported high concerns in relation to having a part-time job throughout their studies, and also with regards to finding a job following their chosen degree.

This survey was targeted to pupils who had indicated that they were studying a Computing-related degree. This high level of concern suggests that whilst pupils are keeping their future careers in mind when choosing their degree, they may also be concerned about what types of career their degree might lead them to.

4.8 Money

This category consisted of four questions. The mean and standard deviation of each question for both years of collection are given in Table 8 below.

It can be seen in Table 8 that high concerns were reported across questions related to the Money topic, consistent with the previous data collection. Pupils reported higher concerns for general financial issues, and for housing fees, suggesting that the concern is sourcing funding. At the stage when pupils were

Table 8. Money: M & SD

Concern	Year 1		Year 2	
	M	SD	M	SD
General money concerns	1.8	0.87	2.2	0.72
Housing fees	1.8	0.80	2.1	0.82
Course fees	1.6	0.87	1.8	0.93
Applying for funding	1.6	0.87	1.8	0.91

surveyed, they would have already applied for student loans and bursaries for their first year of higher education, but would most likely not have received a decision, thus contributing to the higher levels of concern.

4.9 Social

This category consisted of two questions. The mean and standard deviation of each question for both years of collection are given in Table 9 below.

Table 9. Social: M & SD

Concern	Year 1		Year 2	
	M	SD	M	SD
Ability to make friends	1.4	0.98	1.9	0.92
Peer pressure	0.9	0.84	1.3	1.01

It can be seen in Table 9 that some concern was reported, consistent with the previous year.

The high concern regarding the pupils' ability to make friends in the new environment is not surprising: at the stage of survey, one might conjecture that pupils may be part of an established peer group that has been cultivated over a number of years within their existing communities; with the upcoming transition in mind, this peer group would most likely be changing.

5 Discussion

It can be seen that pupils have reported on a number of concerns across several topics, which will be further compared and contrasted with the data collection from the first year in this section.

5.1 Comparison with Previous Data

Due the fact that this study was a continuation of a similar paper-based survey conducted with data collected from the 2014–15 academic school year (Siegel and Zarb 2016), it is informative to explore the relationships between the previously-collected data. While the previous study uses the same questions, the method differed slightly due to paper-based collection methods.

As such, it was initially predicted that significant comparisons might be reserved until after a future, second iteration of the online survey. However, initial findings were quite striking and showed significant correlation between the two data sets. In fact, it will be shown that there was very strong evidence of a positive correlation between the Year 1 and Year 2 datasets.

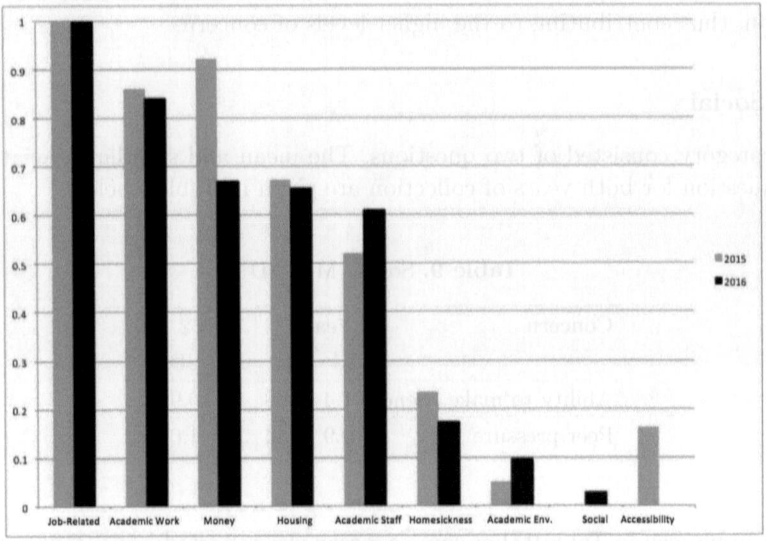

Fig. 1. Topics by rank

Figure 1 demonstrates the ranked order of student concerns by topic, where the ranking is based on the mean response to all questions within each topic area. In the 2014–15 and 2015–16 surveys, the highest-ranking three topics of concern were: Job-Related, Money and Academic Work issues. In both years, jobs were the highest source of concern, both by question and by topic.

Similarly, the bottom three areas of concern were shared across the years. These included concerns in the areas of: Accessibility, Social and Academic Environment. While it is expected that concerns over accessibility support will vary greatly by sample, it is notable that social concerns and those regarding academic environment remain low over the two studies. It is important to note that whilst these are low on reported concern, these are two of the topics that universities might typically consider to be of most concern to students and, as such, often aim to support or target through specialised departments or programmes.

Figure 2 gives the percentile ranking of student concerns by question. The ranking is once again based on the mean response obtained for each question. While there are certainly nuanced differences between the rankings, a notable correlation exists. For these data sets, Spearman's rank correlation coefficient (ρ) can be obtained to understand the similarity of the rankings of the responses by question. This coefficient is given by the equation $\rho = 1 - \frac{6 \sum d^2}{n(n^2-1)}$, where d is the difference in ranks between each of the years and, in this case n represents the number questions, 28. For these data sets, $\rho = 0.903$, demonstrating very strong evidence of positive association between the data. That is, the rankings are very similar by question.

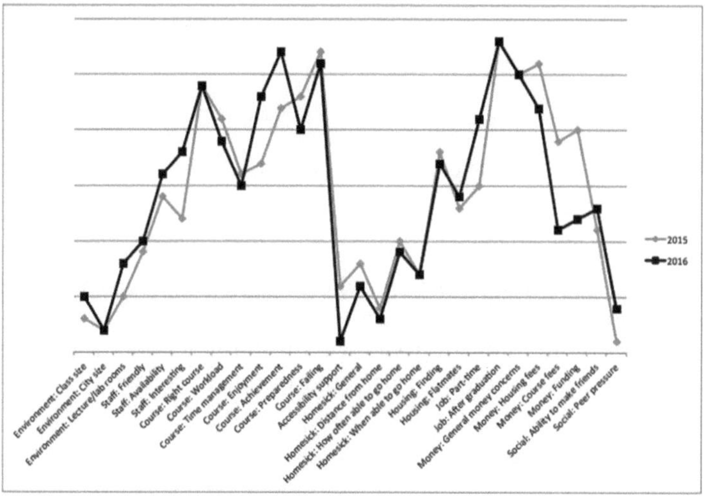

Fig. 2. Questions by rank

Furthermore, we can look to the population Pearson correlation coefficient ($\rho_{X,Y}$) as a measure of the strength of the linear association between the two years of (unranked) data. This will help describe the degree to which the two variables are related and, specifically, how the two variables vary together. For data sets, X and Y, in our case the data from years 2014–15 and 2015–16, respectively, the population Pearson correlation coefficient is given by the equation $\rho_{X,Y} = \frac{cov(X,Y)}{\sigma_X \sigma_Y}$, where $cov(X, Y)$ is the covariance of the sets and σ_X and σ_Y represent the standard deviation of X and Y, respectively. For the collected data, $\sigma_X = 0.359$ and $\sigma_Y = 0.443$ and $cov(X, Y) = 0.145$. As such, the data has a population Pearson correlation coefficient of 0.910, which demonstrates a very strong evidence of a positive correlation. That is, when a value was high in 2014–15, there was a strong trend to also be elevated in 2015–16. Looking at a X-Y plot of the two years in Fig. 3, we can see that this is, in fact, the case.

Whilst the data collected in 2015–16 shows higher concerns than the data in 2014–15, the authors posit that this is due to the fact that all data collection

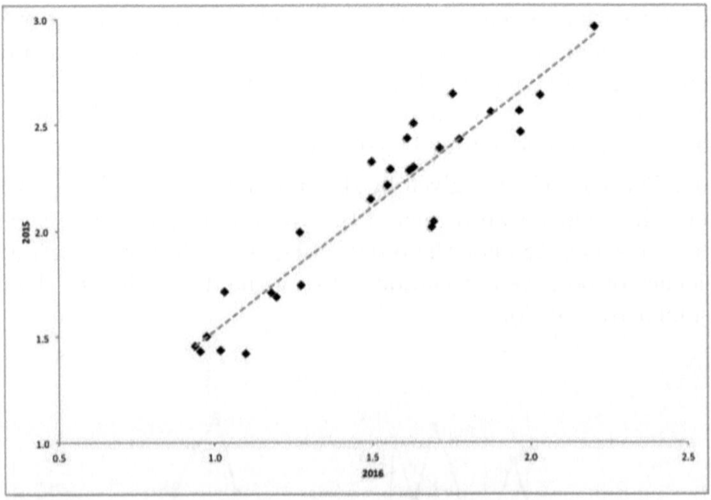

Fig. 3. Year comparison

was done online, allowing pupils to be more honest about their concerns than in the previous set of data collection, where they would typically be faced by academics and peers while completing the survey.

Further collection from future years will establish whether the patterns that are emerging at this stage can be considered trends in concern.

5.2 Recommendations

There are a number of recommendations that can be made based on these results. First and foremost, it is important that these concerns are introduced and discussed: Yorke (2000) shows that students entering higher education had to deal with issues that they had not previously considered. Whilst pupils may not be able to articulate their concerns, or may not feel comfortable doing so, it is important for both schools and universities to introduce these concerns and put this discussion on the table: this stops students from potentially feeling isolated.

Whilst data has been collected and analysed, it needs to be put into context, and an understanding of how to best support these concerns needs to be developed. Within the School of Computing Science and Digital Media at the Robert Gordon University, the work and results discussed in this chapter have been used to inform the creation of a new induction programme for all first year students. This induction aims to act as a buffer between their secondary school studies and their university life, and includes discussions on topics such as time management, independent learning, teamwork and pastoral concern. Furthermore, visualisations of the data have been created and prominently displayed in student spaces to allow as a kickstarter for discussion, both within peer groups and with academic tutors. Student reactions to this programme have been positive, and future work will consider whether this programme alleviates some of the concerns discussed in this chapter. If it does, versions of this programme could be rolled out as necessary to promote the accessibility of higher education: for example, as posters in local communities, guidelines for distance learning, or workshops delivered at schools.

It is possible to distill a number of suggestions for better practice from this analysis, as follows. The following list of recommendations is by no means exhaustive, and is posited by the authors following multiple discussions with prospective and actual students about contextualizing the gathered data:

- Many of the reported concerns occur due to lack of knowledge on part of the applicants: issues regarding potential careers, sourcing funding and finding accommodation are concerns that may be discussed at open-door events organised by the local higher education institution such as Open Days. Too often, academics use pupil-facing activities such as school visits or community events to talk about their subject areas in detail however, the reported results show that it may be worthwhile to use these activities to open discussion on these other avenues that are of concern to applicants transitioning into higher education.
- It is important to set expectation on workload, time management and independent learning. Whilst this is something that is typically targeted during the first few days of the course, like the recommendation above, it may be worth spending some time to discuss these pre-application during pupil-facing visits to alleviate some of these concerns.
- The creation of a safe space where students can discuss non-academic issues, such as their provision to make friends, or problems encountered with living away from family. Whilst many (if not most) institutions offer a degree of student support, this is often manned by staff outside the academic department, who are often unknown to students. Exposure to these staff in class events, or provision of "known" academic staff during these sessions, may be beneficial to students.

Implications for Distance Learning. The study presented in this chapter was carried out with students considering a transition into a traditional face-to-face learning environment. Despite this, some of the concerns presented here are applicable to distance learning environments - out of the 28 concerns presented in this chapter, only 9 are strictly related to a physical environment. Concerns about finance, staff, course structure and jobs are still valid, and relevant to issues found with distance learning situations.

It is shown that in a distance learning environment there is less support available to transitioning students (Clarà and Barberà 2013); many learners feel lost, and are not given any direction or support (Kop 2011; Mackness et al. 2010). Mitigations based on the relevant concerns presented in this chapter could be therefore built into the curriculum to further assist these students.

6 Conclusions and Future Work

Over the data collection periods for this study, it can be seen that pupils continue to demonstrate concerns regarding topics related to money, jobs and course achievement over topics that are considered more traditional, such as academic environment or social issues. The consistency between relative areas of concern over the two years is striking, further suggesting that better understanding of these issues might help schools and universities to better support this group of students.

As further surveys are issued in the coming years, it will be possible to establish whether the current patterns that are presented in this paper could be considered trends in concern. Furthermore, it may be possible to undertake a statistically significant comparison between years for schools who repeatedly take part. This would allow for a detailed view of the data to hone in and understand whether there are any specific issues that are occurring which may be linked by factors such as rurality or socioeconomic status.

References

Askham, P.: Context and identity: exploring adult learners' experiences of higher education. J. Furth. High. Educ. **32**(1), 85–97 (2008)

Clarà, M., Barberà, E.: Learning online: massive open online courses (MOOCs), connectivism, and cultural psychology. Distance Educ. **34**(1), 129–136 (2013)

HESA: UK Performance Indicators; Non-Continuation Rates (2016). Accessed 15 June 2017

Hulme, J.A.: De Wilde, J.: Tackling transition in STEM disciplines: supporting the science, technology, engineering and mathematics student journey into higher education in England and Wales. In: The Higher Education Academy (2015)

Kinnunen, P., Malmi, L.: Why students drop out CS1 course? In: Proceedings of the Second International Workshop on Computing Education Research, pp. 97–108. ACM (2006)

Kop, R.: The challenges to connectivist learning on open online networks: learning experiences during a massive open online course. Int. Rev. Res. Open Distrib. Learn. **12**(3), 19–38 (2011)

Krause, K.-L., Coates, H.: Students' engagement in firstyear university. Assess. Eval. High. Educ. **33**(5), 493–505 (2008)

Leese, M.: Bridging the gap: supporting student transitions into higher education. J. Furth. High. Educ. **34**(2), 239–251 (2010)

Lowe, H., Cook, A.: Mind the gap: are students prepared for higher education? J. Furth. High. Educ. **27**(1), 53–76 (2003)

Mackness, J., Mak, S., Williams, R.: The ideals and reality of participating in a MOOC (2010)

Ozga, J., Sukhnandan, L.: Undergraduate non-completion: developing an explanatory model. High. Educ. Q. **52**(3), 316–333 (1998)

Siegel, A.A., Zarb, M.: Student concerns regarding transition into higher education CS. In: Proceedings of the 2016 ACM Conference on Innovation and Technology in Computer Science Education, pp. 23–28. ACM (2016)

Yorke, M.: Smoothing the transition into higher education: what can be learned from student non-completion. J. Inst. Res. **9**(1), 35–47 (2000)

Higher Education in the Amazon: Challenges and Initiatives

Elaine H. T. Oliveira(✉), José R. H. Carvalho, Horácio A. B. F. Oliveira, Bruno F. Gadelha, Ketlen T. Lucena, David B. Ramos, Ilmara M. M. Ramos, Priscilla B. Nascimento, and Irene Andrea V. A. Carvalho

Institute of Computing, Federal University of Amazonas, Manaus, Amazonas, Brazil
elaine@icomp.ufam.edu.br

Abstract. In this paper, we present an overview of the results achieved by a few applications of educational research towards overcoming the difficulties in the Amazon region's educational scenarios for a better higher education. Manaus, the capital of the Brazilian state of Amazonas, demands an urgent increase in the number of qualified professionals, mostly for its Industrial Pole. However, Amazonas has a poor record of educational performance in all scholastic grades. The situation is especially critical in the rural areas, where the population, separated by miles of uncharted jungle, is usually only accessible through the rivers. This geographic isolation transforms the simplest educational activities into a daily challenge. A sustainable solution to the lack of professionals needs the ability to reach the students with learning activities wherever they are. In this work we share experiences in two different scenarios: the Capital and the rural areas. For the capital, the authors began a large-scale program to address three aspects: (i) reducing student dropouts; (ii) enlarging the reachable community; (iii) offering different levels of knowledge. Four different learning activities were prepared: undergraduate and graduate classes, a talent development program, short-term events and intensive training. For the rural towns, the authors describe the educational experience with undergraduate and graduate distance learning and the ongoing research using learning analytics and adaptive models based on the LMS Moodle. The results are encouraging, as the measurement of the positive impacts of the reported initiative clearly indicates that a structured program would bring solid and long lasting benefits. It is also worth noticing that the proposed solutions may also fit other learning scenarios.

Keywords: Higher education · Involving activities · Distance learning

1 Introduction

The state of Amazonas, the largest state of Brazil, is characterized by a low population density, a common aspect of the entire Amazon region. Amazonas has only 62 municipal districts with 50% of its total population concentrated in just one big city, the two million people capital city, Manaus. The other part of the population is distributed

A. I. Cristea et al. (Eds.): HEFA 2017, CCIS 832, pp. 17–31, 2018.
https://doi.org/10.1007/978-3-319-97934-2_2

in small and less economically developed towns. In this paper, we present both the capital and the country towns' reality, their challenges and perspectives, in an attempt to give an overview of the expected impact of our educational actions.

Manaus hosts the PIM (Manaus Industrial Pole, from the Portuguese), one of the largest Brazilian industrial poles of consumer electronics, information technology and telecommunication sectors. A usual complaint of companies with facilities at the PIM is the lack of highly qualified professionals. Indeed, the job description of open positions contrasts with the professional profile of alumni from the majority of local higher education institutions.

To address this issue, the Institute of Computing (IComp) of the Federal University of Amazonas (UFAM), in partnership with the local subsidiary of a Korean conglomerate, started a large-scale program denominated Promobile, based on the concept of "multiple vortexes of know-how". Vortex is in essence a learning action, working interconnected and coordinated to other vortexes. Four different actions were prepared: undergraduate and graduate classes, a talent development program, short-term events and intensive training. After three years of running program, we proposed a survey to verify the results so far achieved. In this work we will present the program, the results of the survey, and propose corrective actions suggested by the respondents.

Moving to the educational actions for the country towns, most of them suffer from severe geographic isolation. Communities are scattered in the forest and, sometimes, the access is only by boat in the Amazon intricate system of rivers. With all these difficulties, to provide good education to this population constitutes a real and daily challenge. Aiming to bring an opportunity to access higher education to the youth of those remote communities, UFAM through its Center for Distance Education (CDE), created distance learning undergraduate and graduate courses. This CDE project consisted of organizing educational poles, to host courses in Administration, Public Administration, Fine Arts, Biology, Agricultural Sciences and Physical Education. In this work, we will describe this educational experience and present the technology based structure and its pedagogical model that made this proposal a reality. The innovation of this model is to allow the 1,618 students, distributed in 17 different poles, assisted by CDE, to keep pace with their course through a structure of logistical and technological support adapted to their reality. Resources offered by a Learning Management System (LMS), called Moodle, tutors and other specialized tools that support offline activities make it possible to reach the most remote regions of the Amazon with higher education.

The details of the aforementioned initiatives are presented in the following sections. Therefore, this is the structure of this paper. In next section, we present the Promobile program and the results achieved with it. In Sect. 3, the Center for Distance Education and the pedagogical model is proposed. The plugins that were developed and the research being conducted are in Sect. 4. Finally, some concluding remarks are presented in the last section.

2 The Promobile Program

Manaus, the capital of the largest Brazilian state – Amazonas, has approximately two million inhabitants, equivalent to 50% of the population and 80% of the GDP (Gross Domestic Product) of the whole state. The economy is strongly dependent on incentive policies for the industry sector. One of those incentives, named Informatics Law, states that the signatory companies have to invest part of their revenue on research and development (R&D). The Informatics Law has a specific version for industries with facilities in the Manaus Industrial Pole, with the intention to establish a dense value chain of product developed in the region, with local added-value.

In this case, a critical success factor in such a scenario is the availability of highly qualified professionals, from technicians to PhDs. The problem is that the Northern region of Brazil has one of the poorest countrywide performances of fundamental and high schools' students. Moreover, Manaus has only four undergraduate computing courses, and only one graduate course, with an average of 40 to 50 new bachelors and 10 to 15 new MSc and Doctors per year. Therefore, qualified professionals are scarce in the Northern region, especially when considering professionals with an R&D profile in a country with high dropout rates in undergraduate courses of computing [1]. Furthermore, Manaus is not attractive to professionals from other Brazilian regions. The city has problems of services (e.g. public transportation, medical, and education), high living costs, a hot and humid weather typical of the Amazon rain forest and it is isolated from the rest of the country, with connection by road to very few cities.

A group of researchers and company professionals met together at IComp, in UFAM, to find a sound strategy do address the following problem: how to develop long term and sustainable critical mass of R&D profile professionals in a relatively short time frame in Manaus. The strategy is depicted in the next sections.

2.1 Promobile Fundamentals: The Vortex Approach

The proposed program fundamentals lay down on an approach we denominated "multiple vortexes of know-how". Each vortex is in essence a learning activity, working on aspects of qualification (quantity and/or quality). The output of a given activity is directed to the next, benefiting from the motivation of increasing the desire for further knowledge, i.e., the following action offers a new opportunity for the enrolled student. Once finishing a cycle the student will move to a new activity and then to another.

A vortex-based program is conceived to work out specific competences, not to replace the whole curriculum. Given the target competences, we designed Promobile on top of four vortexes, briefly described next:

- Vortex 1 was initially defined as a set of subjects included in the computer science curriculum as optional courses. We prepared a catalog of four classes: two undergraduate and two graduates. Each class has a number of incentives to students to promote participation and performance. Such incentives include open ranking, prizes and the offering of research scholarships, which are the best available scholarships in the university. In order not to jeopardize the other disciplines in the curriculum, to be eligible to receive a prize, the students are supposed not to fail in any class, even if he is top ranked.

- Vortex 2, the New Talents Development program, was defined as participation in research projects. The students received scholarships, with clear granting and maintenance requirements. The candidates enrolled in cycles of off-class project activities with different learning focuses and degrees of difficulties. Students with distinctive performance received leadership coaching (along with higher scholarships). Vortex 2 is also the chosen one to host the soft skills development activities, as the project environment naturally promotes communication and writing skills.
- Vortex 3 comprises a series of short-term events, implemented at the beginning and at the end of each school semester. In the beginning event, the activities are focused on motivation speeches, workshops and hackathon. The ending event has the purpose of evaluate and praise the best projects and students. Additionally, this vortex has the important objective to attract the community to the Program and more participants to enroll in future vortexes.
- Vortex 4 was conceived to enroll the participant in a deep learning experience. Defined as a project-oriented activity, the Advanced Training vortex aims to bring the participant to a point as close as possible to the industry's desired profile. The vortex is applied in a yearly cycle, and accomplished through four modules distributed on six months. Target participants of Vortex 4 are senior students, or alumni, of different universities.

2.2 Program Results in Numbers

Promobile was held in three different cities and started on August 2013 in the city of Manaus, on October 2013 in the city of Itacoatiara, and on November 2013 in the city of Boa Vista. The multiple vortexes provided fast involvement of several sophomore and junior students of undergraduate and graduate levels.

The number of students enrolled in the classes exceeded all the expectations. In total, we had 24 different classes from 12 different disciplines during the three years of the program. Approximately, 650 students were enrolled in the classes, although the number of distinct students may be lower since a student could participate in more than one class.

The main common theme among all the courses was 'mobile devices', since it was one of the goals of the Promobile program. Some courses focused on teaching the required general skills for mobile devices programming such as the disciplines: "Programming Techniques", "Programming Laboratory" and "Programming Mobile Devices with HTML5".

For each of the 24 offered classes, PROMOBILE program awarded scholarship prizes of R$ 1000 (~US$ 280) for the best students. Depending on the performance of the students, one to six prizes were awarded for each class, making a total of 144 prizes awarded during the three years program.

The results of the Vortex 2 were also encouraging. We finished the ramp-up of the New Talents Development Program with the following figures in the three campuses: five doctorate students, 13 MSc students and 33 undergraduate students. Besides being enrolled on research activities, which complexity depends on their level, the students are also encouraged to help with the organization of events, to tutor other students and to present their work in internal meetings. To objectively work on the soft skills and to benefit from the high involvement of the students, we developed an Agile like software development process [2]. The goal is to provide a process to coach the students, while bringing to them an environment close to a systematic development process.

Concerning the short term event (Vortex 3), similarly to what was done to the Promobile program, we created the brand EPA!, from the Portuguese: Encontro de Projetos em Ambientes !nterativos ("!" is "I" as in Interactive, upside down), or Meeting of Projects in Interactive Environments [3] in a free translation to English. The principle behind EPA! is to provide a moment where professionals from different areas would meet and interact together to achieve a common goal, related to a specific theme.

In total we had four EPA! events, reaching up to 1000 registered participants, and eight Android Application showcases, resulting in up to 120 apps. EPA! is a thematic event, e.g. the first one was on Education, and the last one on Entrepreneurship. Each EPA! closes with a hackathon, usually related to conceive, design and deliver a full featured mobile app. In the last EPA!, the hackathon was focused on start-ups. The competitors had to prepare and defend a technology based business.

For the Vortex 4, the Advanced Training on Test Automation for Software Development on Mobile Platforms, we reached 25 subscriptions (the maximum for this activity) in just few days of advertisement via email and social networks. There were four companies represented, besides a group of seven IComp students.

The training was presented on four modules of one week, 4 h per day (20 h per week), from January to May 2014. In between the modules, the students received a workload of around 12 to 20 h to be delivered in the upcoming module. The purpose was to enable student to conciliate the training with their daily activity.

Table 1 provides another perspective on the overall figures. It shows the results depending on the degree of involvement: attendance to one of the events (to promote engagement), students who were exercised mainly on the hard skills, and students who had a full profile qualification. There was no expectation that anyone would be qualified after three days of event. However, the short-event vortex had a crucial role in the entire program. This vortex is responsible to promote the encompassed technologies, to encourage students to leverage their skills, and to motivate more students to be enrolled. This is the most important "engaging" vortex.

Table 1. Promobile: overall numbers for all vortexes

Aspect of qualification	Students (approx.)	From vortex
Engagement	+1000	3
Hard skills mainly	650	1
Hard and soft skills	85	2 and 4

2.3 Further Discussions

After three years of running the program, the professors proposed a survey to verify the results achieved so far, more specifically on training in classroom, and enrolling students in initiation or graduate projects. The survey was designed to answer two questions: (A) How mature is the participant in understanding his career; (B) How is the participant's perception of the program efficiency in meeting his expectations. The survey was composed of 41 yes/no questions covering five different topics: (1) Perception about career and market demands; (2) Information about the program purpose and objectives; (3) Changes in the career plan and expectations of the future; (4) Perceived relevance of the program; (5) Strength and weaknesses of the running program.

Some answers provided interesting insights about the program impact, highlighted in the sequence.

Professional Profile. All students (100%) had a better perception of his course and career than when they started, 100% were fully identified with the career of informatics, and 96.6% knew what they were doing when they chose the computing career, 96.6% would choose computing again, and another 96.6% considered starting a graduate program in computing. But, 27.6% intended to try a public service career opportunity not related with informatics. The reader might think this is a contradiction, but it is not. In a country struggling with high unemployment rates, and an unprecedented economics crisis, the stability of a public service job will always be an attractive option. It is interesting to notice that for 82.8% the Program had a positive influence on their career decision.

Most of them (90.7%) agreed that English proficiency as a foreign language is important for their career; however, only 41.4% affirmed they can sustain an English conversation via Skype. Only 48.3% reads more than 300 pages of books per year and 18% confessed severe issues with writing skills in Portuguese (our native language).

These figures are not new to anyone involved with higher education in Brazil. And they just reinforce the importance of learning programs like the proposed one.

Program Relevance. All respondents (100%) agreed that the Program objectives do not conflict with the official curriculum objective, which is correct. However, 75% think that Promobile was created to provide professionals to the private company that funded it, which, technically, would conflict with the curriculum objective to prepare professionals to the market, and not to a specific company. This is something that we will have to clarify among the students: Promobile is indeed an academy-industry partnership with the purpose to promote critical mass on a specific know-how. Any company can benefit from it, including, of course, the sponsor.

Most of them (96.6%) agreed that Promobile was somehow beneficial to their professional skills and 89.7% said they are better prepared to the market after participating in the program. For 79.3%, their employability improved due to Promobile, and 79.3% would participate again even for free (Promobile scholarships were the best ones available at the University). These are the most indicative statements about the relevance of Promobile. For 34.5%, anyone out of the Program will necessarily have lower employability.

Most of them (96.7%) would take Promobile to other universities in our region, 89.7% would like to see others researchers as part of the Program (currently we have eight out of 25 researchers), and 86.2% would like to see the Program promoting others disciplines in the curriculum.

These figures confirms the statistics presented before, that the Program was considered relevant to the participants, and they agreed to change so that it will reach other universities, research areas and course disciplines.

With this, we conclude this section and the main discussion about our capital, Manaus. To see more detailed information on Promobile program, please read [4–6]. Next, we will talk a little about distance education strategies to inner towns.

3 Center for Distance Education

Distance Education (DE) is a teaching modality, applied to any level of education, in order to extend the possibilities of access to knowledge, democratizing access to education. In DE, according to [7], "students and teachers are in different locations during all or most of the time they learn and teach", providing benefits that create more flexibility to the student in terms of time and place of study.

The Center for Distance Education (CDE) [8] is a supplementary unit of UFAM. CDE is responsible for providing initial and continuous education in undergraduate and graduate courses to many cities in the state of Amazonas. These courses are offered in distance mode and rely on technological and pedagogical support from a well-structured and capable team. This team is responsible for applying information and communication technology in educational process.

CDE offers courses since 2007 and currently has 1,618 students distributed in six undergraduate courses: Administration, Public Administration, Fine Arts, Biology, Agricultural Sciences and Physical Education. Besides those, the postgraduate courses are: Production of Teaching Materials for Distance Education, Public Administration, Municipal Public Management and Health Administration. All courses are free.

The undergraduate courses are distributed in 17 different poles. Amazonas is the largest Brazilian state, and the 9th largest country subdivision in the world. Its area is larger than Germany, France, United Kingdom and Japan all together.

Most cities suffer from severe geographic isolation, as they are scattered throughout the forest and its only access is by river. This situation makes population access to basic services, including Education, a difficult task [9].

3.1 Pedagogical Model Supported by Technology (PMT)

According to [10], a pedagogical model adapted for distance education should seek for new paradigms in teacher-student relationship. Pedagogical, organizational and technological elements of this model should be harmonized so that learning happens interactively, collaboratively and autonomously.

CDE has a differential model of distance education which adapts to the reality of the poles to which undergraduate courses are offered. As large geographical distances from its headquarters in Manaus and precariousness of Internet access connections are daily difficulties to these places, the pedagogical design of courses is mediated by resources such as printed material, videos and a Learning Management System (LMS). Thus, the model of distance education implemented by CDE – called Pedagogical Model supported by Technology (PMT) – is composed by technological resources supported by a qualified staff.

One of the resources is the printed material which preparation follows strict quality standards, including special care against plagiarism. This printed material consists of several books, each with an ISBN number. Once the books are ready they become available to the public in general.

In CDE, each term is called 'module'. In each module the student uses at most three books. Each book covers at most three disciplines, varying from course to course. An instructional manual is also available, which explains how the books are organized and the meaning of the icons used. The videos are produced and developed seeking to provide the student with an environment similar to a traditional classroom [11]. The LMS of CDE is based on Moodle [12] (versions 1.9 and 2.0), and each course has a unique database that is stored on servers, whose technical support is responsibility of the Center for Information and Communication Technology at UFAM.

All poles have a computer lab, a library and a boardroom. Also, each pole receives an "access kit" composed of a satellite dish for Internet access, 30 computers with Linux operating system (educational version), a laser printer and a wireless router. It is worth mentioning that Internet connection at the poles is only through satellite, at maximum speed of 512 Kbps, with an average typically of 100 Kbps. Thus, adapting to this reality, a peculiarity of the scenario is to avoid activities that require online participation (such as chats) or downloads of videos or files with many images. When an activity requires supplementary notes or videos, DVDs and/or printed material are sent via conventional mail.

With these resources, the pedagogical model works like this: each pole has a local coordinator who is responsible for its infrastructure and maintenance. Students of a course are divided into groups of 25, guided by face-to-face tutors, whose functions are to provide pedagogical and technological support. There are also online tutors who are experts in the course and in charge of supervision and correction of the activities of each discipline.

Coordinators of tutors and course coordinators stay at the headquarters in Manaus. Teacher's role is divided into two types: author and lecturer. Author teachers work out the books and organize the courses in the so-called "Virtual Classrooms", using the LMS. Lecturer teachers have the task of monitoring the development of the course. As the teams are independent, it is possible to have the same teacher in both teams.

We can consider that the PMT of CDE has two divisions: the first one consists of a traditional classroom, which we call "face-to-face mode", while the second division is distance education, supported by LMS, called "virtual mode". The face-to-face part of the courses consists of the daily support of the local tutor and a face-to-face class every 90 days; taught by the lecturer teacher responsible for the course or by other assistant teachers, from the headquarters in Manaus. This class is called "Disciplinary Introduction" when the teacher presents the course (or discipline), its content, the supporting material and activities. Moreover, in this opportunity the first evaluation activity of the course is held.

On the other hand, the virtual part fully occurs in the LMS. It plays a key role in the progress of the course, because it is where the student follows the content, performs the suggested activities, and interacts with peers, teacher and tutors. This virtual environment encourages the practice of collaborative learning as well as individual and collective construction of knowledge. All course curricula are organized in the LMS, where disciplines are available, according to the offerings of the school year. The student is enrolled in the class and follows the content through presentation files and supplementary texts. The evaluation activities consist of research and questionnaires. A tool commonly used in CDE LMS is 'Forum' where the student has the opportunity to interact with peers, tutors and teachers, providing feedback on any issue established by the teacher.

For more information on CDE and its strategies, refer to [13]. As the LMS is a key point in PMT, our research group seeks for new findings and ways to support students and teachers in the learning process. Next section presents some tools we have been developing and some initial findings.

4 Learning Analytics Tools and Adaptive Models

Since 2013, our research group has been developing learning analytics tools and adaptive models based on the LMS Moodle. Moodle is a worldwide adopted learning environment. It is an open-source platform supported by a global community. The standard version comes with the all the basic functionalities needed for the management of online courses. However, for extra support such as learning analytics, we need to look for the available plugins in Moodle site or develop our own. After some browsing and research, although, there are 1450 plugins[1], with various purposes, we could not find specific ones that match our research and application interests. So we have decided to develop our own and try to contribute to the community. This would give us some experience on personalization and adaptation, desirable for our research group. The idea is to make the plugins available to teachers and students in attempt to help both distance and blended education. All the plugins are in beta version and need extensive tests. In this paper we will give an overview of four plugins that we have developed. Further details may be found in the references indicated in each corresponding section. The first one is MobiMonitor, presented next.

4.1 MobiMonitor

MobiMonitor [14] is a mobile app integrated with a multi-agent system, which main goal is to instrument course mediators with tools for monitoring student participation in assignments and forums. It also provides support to identify students who need a direct pedagogical intervention in order to avoid possible course evasion, failure or dropout. Instead of analyzing student's performance by the end of the current discipline, MobiMonitor enables mediators to check these data on the fly, classifying students'

[1] https://moodle.org/plugins/.

participation into four levels (excellent, good, low and very low) as shown in Fig. 1. The mediators use mobile applications to access this report and to send alert messages or warnings to students, according to their attendance status.

The first initial validation of MobiMonitor was done in two courses of Basic Computing. During use, students' performance in activities and the amount of posts in the forum in poles of the Santa Isabel and Boa Vista were collected. Although these data were collected at the end of the course, it could be done anytime at the mediator request. We could infer, from the data, that most students under-performed in both assignments and forums. Surprisingly, from our experiment sample, none of the students showed an excellent participation.

With this application, the mediator can monitor the whole class, pay close attention on those students who have not a good or excellent participation and try to motivate students through constructive messages.

As mobile devices screens are usually size limited, we also have plugins that can be used in desktops or notebooks that may allow further and more detailed learning analysis. Next, we will present LMS Monitor, which helps performance analysis (Fig. 2).

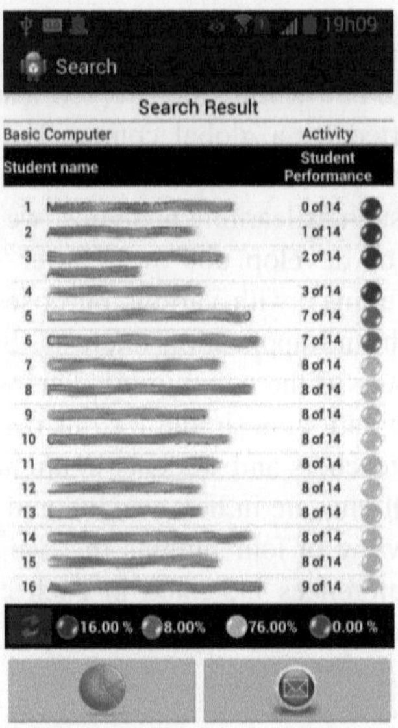

Fig. 1. MobiMonitor – student performance

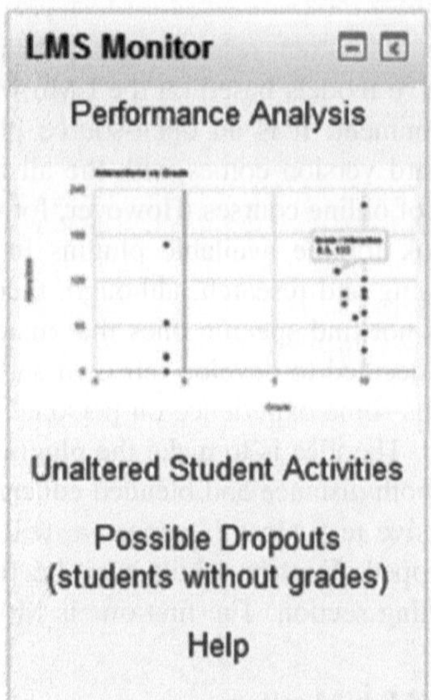

Fig. 2. LMS Monitor – performance analysis

4.2 LMS Monitor

As showed in previous figure (Fig. 2), LMS Monitor [15] is another learning analytics tool that aid educators to analyze students' academic performance in LMS. It provides resources to improve integration and information needs to assist professors or mediators of a virtual classroom by using graphic techniques of information visualization: performance analysis and progress, possible dropouts, student's activities and resources tracking, among others. In addition to support mediation, LMS Monitor also allows students to self-assess themselves according to their commitment in course activities.

The amount of interactions that each student establishes with the elements in the LMS can represent his behavior. As no threshold of access quantities or element interactions has been established, observing this indicator can help in identifying some "risky" behavior when it is outside the (average) pattern of interactions.

The interaction indicator may be out of the average in two opposing situations: the reduced number or the excessive interactions established with one or more LMS elements. In the first case, the low interaction with the elements of the environment can compromise the learning process due to the lack of contact with content made available through these elements. In the second case, a high number of interactions with one or more resources may reveal that the student is experiencing difficulties in manipulating these elements in the LMS. In a preliminary analysis of the data, it was possible to establish the correspondence between the dropout and failure rates and the average of interactions in course activities. Another information to consider is that the dropout rate in one of the classes, the only one to use the mobile application, was the lowest among all studied. It is true that we cannot yet conclude that the use of this application motivated the students, but it shows itself as a point of observation and experiment with future classes.

LMS Monitor seems to be a useful tool to support performance management in an LMS, by monitoring students' participation in assessment tasks. From the data collected it is possible to identify possible dropout or failure, and also to provide subsidies for a direct intervention to avoid these problems. In addition, this application presented an intuitive monitoring of the activities for teachers and students the proposal, due to the use of graphical information visualization techniques. Further details on this tool can be seen at [16].

4.3 LPGraph

The third plugin we developed was LPGraph [17], a tool for identification and representation of learning paths that shows a visualization of actions performed in the LMS. The idea is that it shall help teacher and students to adapt both navigation and content in virtual learning environments. LPGraph model extracts the data stored in database records and uses the graph structures as a basis to model the information about the students learning paths and to present them visually to teachers and tutors. The graph vertices represent the resources and activities from LMS course and the edges indicate the path performed by student while he/she is interacting with the resources and activities.

The proposed tool was applied in two courses during one academic semester. It seemed useful in monitoring learners by a visual tool that aids the behavioral analysis through learning paths. The plugin was used in two classes: Introduction to Computer Science, with one class to the undergraduate majoring in Physics (28 students) and Electrical Engineering (42 students), and Discrete Mathematics (43 students) to the majoring in Computer Science. An example of a learning path graph is shown in Fig. 3.

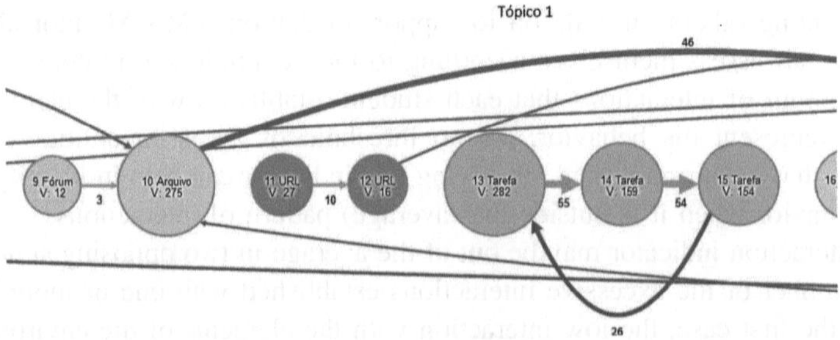

Fig. 3. Part of the learning path graph of Discrete Math to Computer Science

The tool was effective in representing the paths and, thus, to allow the student to follow-up within the LMS. LPGraph does not only benefit distance education, but any teaching modality that uses Moodle or other LMS as a learning environment, communication and interaction among teacher, student and content. The tool can be used in a class in progress or in an already closed class. Real-time analysis allows immediate intervention by the teacher.

The model can also contribute to solve problems such as cold start, when a system needs to make recommendations, but does not yet have data about the users, help groups recommendations to carry out collaborative activities, to follow up student evolution over time, to verify the influence of the available resources with success and failure rates on the disciplines, to associate the student's behavior with dropout tendencies, to verify which paths improve student's performance, and also to contribute to the creation of adaptive LMS. Thus, the presented model can become a new source for analysis and mining of educational data in virtual learning environments.

Besides looking learning paths, it would be interesting to investigate how grouping techniques would work using this approach. This is presented by the next plugin, M-Cluster.

4.4 M-Cluster

In an educational environment there are several activities that require collaboration, so it is often necessary to organize groups of learners to perform these activities. To contribute to the process of teaching and learning, this approach is intended to suggest student groups to accomplish collaborative activities based on learning paths, extracted from Moodle.

The last plugin to be presented is M-Cluster [18], a tool to support group formation for collaborative activities in LMS. The recommendation is based on student's learning paths and uses K-means clustering algorithm with three different similarities metrics.

The first results obtained by the tool were generated from real data from the LMS. The data were from two already closed classes of the discipline Discrete Mathematics, a class of the Computer Science and another of the Computer Engineering course, both at UFAM. The classes, with 44 and 40 students respectively, provide data on learning paths. A system processes this data, transforming it into a knowledge base in which the

K-Means algorithm is applied to generate the initial grouping. It was verified that with the paths it is possible to obtain behavioral patterns to generate the groups of the apprentices.

The first analysis of group formation was done based on a pair activity assigned by the teacher in the environment. This time, we used data from an ongoing Discrete Mathematics discipline, of a blended learning class with 40 students, also of the Computer Science course (see Fig. 4).

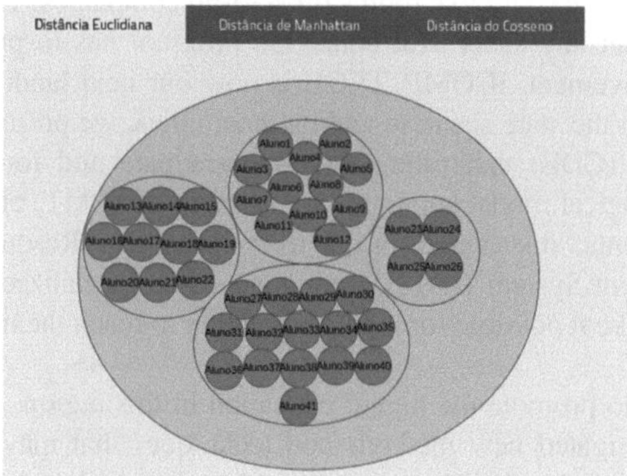

Fig. 4. Groups by K-Means algorithm, using three different distances: Euclidian, Manhattan and Cosine

In Fig. 4, there are 41 students (*aluno*, in Portuguese) – one of them chose not to participate on the research. They are grouped by K-Means algorithm, using three different distances: Euclidian, Manhattan and Cosine. The groups suggested by the tool were mostly with homogeneous characteristics and the groups formed by the students themselves were mostly with heterogeneous characteristics. However, grades from the first configuration were equal or higher in 75% of the cases. This specific observation may indicate that in this case, the best should be to have students with similar behavior. But these are only preliminary results and further research is necessary. Currently, we are analyzing other classes to get to a common denominator about characteristics that will be important for grouping suggestion. The idea is that the type of group depends on the teacher's goals.

With this version of the system it was possible to glimpse the opportunities to improve the learning process, from a better implementation of group formation based on learning paths and machine learning methods.

5 Final Considerations

This paper presented some of the education challenges faced by the Federal University of Amazonas. In the capital, Manaus, we have worked to improve the scenario through a large scale education program, called Promobile. It was a qualification program based on "multiple vortexes of know-how" aiming to generate critical mass in technological human resources. The program has accomplished its goals and showed a relevant influence on the participants' career path. Summing up, more than a thousand students were reached in the program events. About 650 had the opportunity to develop hard skills, mostly in the courses. And 85 students were either participant of new talents development program or advanced training, which main purpose was to develop soft skills. An important legacy of the Promobile program was the newly created ICOMP_TEC initiative, a real-estate composed by a building with three floors, auditorium, a coworking space, and five rooms for start-up companies. The principle behind this program was that any successful education program has to promote a sustained quality of life improvement. ICOMP_TECH is now our next landmark.

Taking a look on the state scenario and its inner cities, we presented the Center for Distance Education (CDE) that offers six undergraduate and four graduate remote courses. The pedagogical model supported by technology (PMT) allows it to keep the current 1,618 students, distributed in 17 different poles. Resources offered by a Learning Management System (LMS), tutors and other specialized tools that support off-line activities make it possible for higher education to reach the most remote regions of the Amazon.

As an initiative to promote the higher education in this region, the involved group of researchers investigated new methods and techniques that may help teachers and students. We have been developing new tools and analyzing data from LMS to support learning analytics and adaptive models. We believe that, with these modest initiatives and others that have been accomplished by other researchers and institutions, the Amazon will succeed in fulfilling the demand of highly qualified professionals entirely by its local higher education institutions.

References

1. Ambrosio, A.P., Almeida, L., Franco, A., Georges, F.: Assessment of self-regulated attitudes and behaviors of introductory programming students. In: Proceedings of IEEE Frontiers in Education Conference (FIE 2012), Seattle, WA (2012)
2. Valentin, E., Carvalho, J.R.H., Barreto, R.: Rapid improvement of students' soft-skills based on an agile-process approach. In: Frontiers in Education Conference (FIE), pp. 1–9. IEEE, El Paso (2015)
3. EPA! Encontro de Projetos em Ambientes Interativos. http://epa.icomp.ufam.edu.br. Accessed 23 June 2017
4. Carvalho, J.R.H., Oliveira, E.H.T., Silva, A.D.: Generation of critical mass in education: an approach based on multiple vortex. In: Proceedings of IEEE Frontiers in Education Conference (FIE), Madrid, vol. 1, pp. 1471–1478 (2014)
5. Carvalho, J.R.H., Oliveira, E.H.T., Carvalho, I.A.V.A.: STEM education program evaluation survey: a report of experience. In: 46th Annual Frontiers in Education Conference, Erie, PA (2016)
6. Oliveira, E.H.T., Oliveira, H.A.B.F., Carvalho, J.R.H.: Generation of critical mass in education: an initiative to engagement. In: 46th Annual Frontiers in Education Conference (FIE), Erie, PA (2016)
7. Moore, M.G., Kearsley, G.: Distance Education: A Systems View. Thompson Learning, São Paulo (2007)

8. CDE - Center for Distance Education. http://www.ced.ufam.edu.br. Accessed 23 June 2017
9. Malvezzi, W., Neles, K., Bressan, G.: Classroom mediated by technology model: a new proposal for education to remote regions of the Amazon. In: 41st ASEE/IEEE Frontiers in Education Conference (2011)
10. Behar, P.A.: Constructing pedagogical models for e-learning. Int. J. Adv. Corp. Learn. (iJAC) **4**, 16–22 (2011)
11. Carvalho, F.C.A.: Tecnologias que educam: ensinar e aprender com tecnologias da informação e comunicação. Pearson Prentice Hall, São Paulo (2010)
12. Moodle - Modular Object-Oriented Dynamic Learning Environment. http://moodle.org/. Accessed 23 June 2017
13. Oliveira, E.H.T., Lucena, K.K.T., Nozawa, E.H., Lucena, W.: Distance education with remote poles: an example from the Amazon region. In: Proceedings of IEEE Frontiers in Education Conference (FIE 2012), Seattle, WA (2012)
14. Lucena, K.K.T., Silva, J.S., Frota, V., Oliveira, E.H.T., Gadelha, B.F.: MobiMonitor: a mobile app for monitoring distance courses in the Amazon region. In: IEEE Frontiers in Education Conference, Madrid, vol. 1, pp. 2126–2133 (2014)
15. Lucena, K.K.T., Silva, J.S., Oliveira, E.H.T.: WebMonitor: uma ferramenta para monitoramento e acompanhamento de cursos em um AVA. In: XXVI Simpósio Brasileiro de Informática na Educação, Maceió, vol. 1, pp. 249–258 (2015)
16. Antunes, O., Lucena, K.K.T., Oliveira, E.H.T., Gusmão, T.H.Y., Nozawa, E.H., Brodbeck, W.: LMS monitor: an application for academic and performance monitoring of students in LMS. In: XXI Congreso Internacional de Informática Educativa (TISE), vol. 12, pp. 45–52. Nuevas Ideas en Informática Educativa, Universidad de Chile, Santiago (2016)
17. Ramos, D.B., Ramos, I.M.M., Nascimento, P.B., Oliveira, E.H.T.: Uma Ferramenta baseada em Grafo para Identificação e Visualização de Trilhas de Aprendizagem. In: XXI Congreso Internacional de Informática Educativa (TISE), vol. 12, pp. 237–243. Nuevas Ideas en Informática Educativa, Universidad de Chile, Santiago (2016)
18. Ramos, D.B., Ramos, I.M.M., Nascimento, P.B., Gadelha, B.F., Oliveira, E.H.T.: Sistema de Combinação Social em um Ambiente Virtual de Aprendizagem. In: XXI Congreso Internacional de Informática Educativa (TISE), vol. 12, pp. 325–332. Nuevas Ideas en Informática Educativa, Universidad de Chile, Santiago (2016)

Computational Methods for Analysis of Language in Graduate and Undergraduate Student Texts

Samuel González-López[1](✉) and Aurelio López-López[2]

[1] Department of Information Technologies and Communication, Technological University of Nogales, 84094 Nogales, Sonora, Mexico
sgonzalez@inaoep.mx
[2] Department of Computational Sciences, National Institute of Astrophysics, Optics and Electronics, 72840 Tonantzintla, Puebla, Mexico
allopez@inaoep.mx

Abstract. Often, academic programs require students to write a thesis or research proposal. The review of such texts is a heavy load, especially at initial stages. Natural Language Processing techniques are employed to mine existing corpora of research proposals and theses to further assess drafts of college students in information technologies and computer science. In this chapter, we focus on examining specific sections of student writings, first seeking for the connection of ideas identifying the pattern of entities. Subsequently, we analyze the justification and conclusions sections, studying features such as the presence of importance in justification and the level of speculative words in a conclusion section. Experiments and results for the different analyses are explained in detail. Each analysis is independent and could allow the student to analyze their text with a set of tools with the aim of improving their writing.

Keywords: Language analysis · Theses writing · Weak sentences
Sequence of concepts

1 Introduction

Research writing is a complex task. In educational institutions, academic programs conclude with the thesis writing. Nevertheless, students can attain academic degrees through alternatives that do not place emphasis on a written work. In this context, the student's ability to write a scientific paper is not of the expected level. A study about the perception of the difficulties of students when writing the discussion section of a thesis [1], that applied depth-interviews to supervisors and students, commented about the uncertainty of students concerning the content that should be included in the discussion section and how it should be organized. This information was surprising to supervisors, considering the time and feedback that students received.

These documents must comply with the drafting characteristics established by institutional guidelines, such as format, orthography, structure, legibility, or argumentation. Our efforts aim to help students in preparing their drafts by examining the

A. I. Cristea et al. (Eds.): HEFA 2017, CCIS 832, pp. 32–51, 2018.
https://doi.org/10.1007/978-3-319-97934-2_3

text of three sections of a thesis: problem statement, justification and conclusions section in a proposal or thesis draft.

These sections of a thesis have the characteristic of having a larger extension than the rest of the elements. Therefore, the automatic or computational identification of attributes in a thesis draft could help students to improve it. Below, we describe the three sections analyzed [2].

Problem Statement: The problem statement formulates a need and sets decisions and the direction of the study to achieve certain objectives, so that relevant data are collected, considering these goals. Some points to consider are:

- Problem statement should be written clearly,
- It should express the relationship between two or more concepts or variables,
- It must involve the possibility of an empirical test.

Justification section: This indicates the reason for the investigation, giving the explanation therefore. Thorough justification should demonstrate that the study is necessary and important.

Conclusion section: a good conclusion must include the following features: an analysis of compliance with the research objectives, a global response to the problem statement, a contrast between results and theoretical framework, future research work and acceptance or rejection of the established hypothesis.

This chapter focuses on examining each section, first seeking for the connection of ideas identifying the pattern of entities. Subsequently, we analyze the justification and conclusions sections analyzing features such as the presence of importance in justification and the level of speculative words in a conclusion section. In Sects. 4 through 6 each of the experiments are explained in detail. Each analysis is independent and could allow the students to examine their text with a set of tools with the aim of improving their writing. In Fig. 1 the main characteristics evaluated in the three sections are presented.

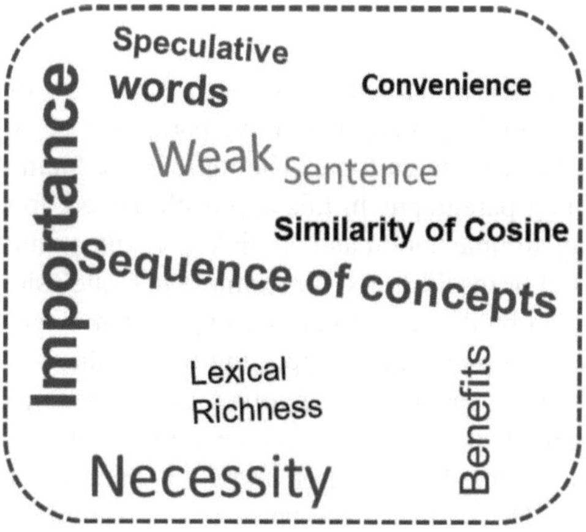

Fig. 1. Main features analyzed in student writing sections

Also, in this chapter, we present a group of Natural Language Processing techniques to mine existing corpora of research proposals and theses to further assess proposal drafts of college students in information technologies and computer science. Furthermore, we intend that these developed methods, implemented in a system, indirectly helps the academic advisor by reducing the time dedicated to the draft review, enabling to focus on content.

The results of analyses reported here are part of a larger project that may help students to evaluate early their drafts and facilitate the reviewing process of the academic advisor. Thus, this chapter mainly shows the results of the application of previously developed methods.

The scope of our work is mainly the modeling of the structure of a thesis (long sections), that is, we focused on the elements that each section requires, for instance, that in a justification the concept of importance is present. Nonetheless, the content is not addressed in depth. We look for our methods to be complementary in the revision of the thesis. The idea is that the academic reviewer does not spend a lot of time on the structural revision but rather focuses on content.

This chapter is structured as follows. The next section reviews previous related research. Section 3 describes the Corpora employed to mine and validate the experiments. Section 4 details Conceptual Sequence Analysis experiments and results. Section 5 analyzes and discusses the Weak Sentences Method. Section 6 shows a Justification method to identify essentials elements. Finally, Sect. 7 details the conclusions and future work.

2 Background

Three topics are central to this research; Conceptual Sequence, Weak Sentences, Essential elements in Justification section. We review previous approaches and related work in each of the topics.

2.1 Conceptual Sequence

A document with an appropriate structure presents a clear flow of topics through their paragraphs. For instance, in the paradigm of a five-paragraph essay [3], the introduction and conclusion share the same main topic. The topic is the theme or subject matter of the essay. The remaining paragraphs in this approach named "body paragraphs" contain details of the essay argumentation and are linked via the main topic. This approach is similar to the proposal or thesis drafts, for example in Conclusions section, where the paragraphs are connected by the same topic, we expect that is the same of the problem statement, and that include paragraphs supporting the results.

The connection between paragraphs involves the interconnection of each of the sentences to the paragraph through its grammar constituents as subject and object. These constituents are observed as a pattern and allow the correct interpretation of the information in the text. The sequence presented in the pattern of the constituents can characterize specific types of discourses and therefore contribute to the assessment of the quality of the generated text [4]. In our work, for example, in the Problem

Statement section, the set of sentences that integrate each paragraph are interconnected by the same central topic. This flow of connections provides an adequate sense of what the student seeks to address in the proposal. Under this approach, our research takes as a working unit the sentence.

In [5], the authors present a related study with the semantic flow of essays of students learning English, and argue that the paragraphs of essays have an internal flow, i.e. each paragraph connects with an adjacent paragraph. If paragraphs are connected in ascending order, the essay is suitable for a semantic approach. They used singular value decomposition (SVD), where each paragraph of the essay was represented in a vector space, and then they took the measure of distance between vectors to determine the semantic proximity of paragraphs. This proximity depended on the topic of each paragraph. However, the topic flow in the essays was small, but present. They concluded that is possible to obtain a better semantic flow on collections of essays where the quality differences are more significant.

In our work, we took advantage of a tool based on the Entity Grid (EGrid) technique to represent discourse, which was proposed to evaluate coherence in text [6]. The technique generates a representation constructed as a two-dimensional array that captures the distribution of entities in discourse across sentences, where rows correspond to the entities of discourse and the columns represent the sentences. The cells can have values such as subject (S), object (O), or neither (X). The main idea of this representation is that if the object and subject are present in the sentences, the assessed coherence is stronger. For instance, given the following three sentences with the main entities of discourse bracketed:

S1: [The Justice Department] S is conducting an [anti-trust trial] O against [Microsoft Corp.] X with [evidence] X that [the company] S is increasingly attempting to crush [competitors] O.
S2: [Microsoft] O is accused of trying to forcefully buy into [markets] X where [its own products] S are not competitive enough to unseat [established brands] O.
S3: [Microsoft] S claims [its tactics] S are commonplace and good economically.

We observe how the entity Microsoft appears initially as X in S1, then as Object in S2, and as Subject in S3, revealing certain coherence. The EGrid technique generates a model which is built from a specific corpus and such model is further used to evaluate new texts. The authors in [6] used 3,991 samples for training and 4,143 were used for testing on the topics of earthquakes and accidents. Similar to our work, we identify the subject and object of sentences with the goal of detecting the sequence of concepts, instead of local coherence. But in contrast, our proposal uses paragraphs to train the model that will assess the concepts of the rest of the section.

2.2 Weak Sentences

Each element of a thesis has a particular structure, as established in guidelines. These guidelines are established by educational institutions and allow the sections to differentiate each other. For example, a general objective is different from a hypothesis because the objective expresses an action to be performed globally, while the hypothesis is written as an affirmative proposition. Thus, a sentence has characteristics that fit into a particular section. In our approach, a weak sentence is a statement that does not follow these guidelines for the conclusion section.

The conclusions provide a global answer to the research question and contemplate the following aspects:

- Analysis of compliance of each of the goals of the research.
- Acceptance or rejection of the hypothesis.
- The contrast between the foundations and results, analyzing each paragraph of foundations and commenting on the results.
- The constraints that guided the investigation.

Therefore, if a sentence is identified as weak, it can be improved by the student to fit in the conclusions section. Our efforts to identify a weak sentence are focused in the conclusions section, since it has feasible characteristics of being processed computationally. The proposed method identifies weaknesses in sentences, such as the use of general instead of specific terms, or the absence of reflections and personal opinions. For instance:

In the project, we have developed two concept tests, one has been to do the survey that collected data from people, and another has been to make a concept test of the peripheral.

Here, it can be observed that the student describes a part of the experimentation, instead of providing a value judgment of the results. This sentence would be more suitable to another section of the thesis, for example, in methodology.

This kind of analysis has been done through systems that monitor the student and send recommendations. The tutor Writing Pal [7] offers strategy instruction and game-based practice for developing writers, and was also used with an experimental group, obtaining significant improvements in students who participated in the experiment. Our work considers such lexical, cohesion and rhetorical features, augmented with other measures with the goal of capturing behaviors specific to the conclusions section.

A dialogue-based ITS called Guru has an animated tutor agent engaging the student in a collaborative conversation that references a hypermedia workspace, displaying animating images significant to the conversation [8]. Another dialogue-based ITS, Auto Tutor, uses dialogues as the main learning activity [9]. One interactive essay-writing tutor was designed to improve science knowledge by analyzing student essays for misconceptions. The authors presented five modules that allow the system to identify if the student is using concepts improperly [10]. We take some inspiration from their work in developing our model of weak sentences. All these works used natural language to interact with the student, as we do.

2.3 Essential Elements in Justification Section

We explore the presence of four essential elements in the justification section [2]: importance, necessity, convenience, and benefits. The justification section indicates the reasons to achieve the goal stated in the proposal or thesis. Below the elements are defined using questions that the student should answer when writing such section:

Importance: What is transcendence of the research?
Necessity: What are the essential aspects to be covered by the investigation?
Convenience: What is the purpose of the research?
Benefits: What are the enhancements obtained from the research?

In some cases, the essential elements can appear at distinct levels in the paragraphs that make up a justification and can be visualized as four core dimensions.

Our method aims to find evidence of these dimensions in the Justification. Below, we provide a couple of examples of justifications that emphasizes the "importance" and "convenience" elements:

Currently, the University has approximately 3000 computers that need to be in perfect condition and must be used by the staff of this institution. It is important to use technology to help optimize resources, increase productivity, and improve the performance of support department.

The underlined sentence reveals the importance of the research, where the phrase "increase productivity" is related to the transcendence of the investigation. While the following paragraph serves to illustrate convenience:

Due to the increase in resolution of current images, object recognition requires higher processing, where the power of multiple processors or multiple cores can greatly reduce the execution time, that in the case of traditional sequential processing cannot be achieved. With the power of parallel processing, many processes can be fed to different processing units, resulting in a substantial improvement in runtimes, therefore, this is seen as a solution for the best performance of algorithms that require high computing power.

In the underlined text, the main purpose of the approach is provided. Similarly, the other essential elements can appear in a justification.

In [11] Burstein proposed a machine learning approach to identify the thesis (main topic) and the conclusion section in student essays written on different topics. The results were favorable to identify the conclusion section, because the conclusion always was located at the end of the essay. Our works assume that the justification section is previously bounded, and the goal is to identify internal expected elements (core dimensions) of such section.

The identification of features in academic writings is similarly found in the study of [12], in which the conclusions are analyzed using lexical features, cosine similarity and speculative terms. In our work, we conducted a characterization of the sentences of justification section to n-grams level and we used a language model to capture some patterns related to the elements.

Moreover, the study of Daunaravicius [13] analyzed papers to identify sentences that do not fit to the scientific writing genre. Some features addressed in the study were a formal notation and specific terminology used in the papers. Similarly, the solution proposed to assess justification section focuses on the structure of each essential element, such as the importance or necessity.

3 Background

The collection was composed of 468 documents: theses and students' research proposals written in Spanish language. Each item of the collected corpus is a document that was evaluated at some point by a reviewing committee. In the corpus, two kinds of documents can be distinguished, graduate level (Doctoral and Master degree) and Undergraduate (Bachelor and TSU degree). Documents come from universities and research centers in México.

To build the collection, complete documents (theses and research proposals) were downloaded from universities' public repositories. Subsequently, from each document the following elements were extracted by hand: Problem Statement, Justification, Research Questions, Hypothesis, Objectives, Methodology and Conclusions section. The following table shows the amount of collected data per study level.

Below, Table 1 details each element extracted from the corpus. The amount displayed is 2,216 elements extracted from the whole corpus.

Table 1. Detailed corpus

Corpus Section	Graduate	Undergraduate	Total
Problem statement	132	100	232
Justification	108	132	240
Research questions	133	19	152
Hypothesis	71	26	97
Objectives	218	212	430
Methodology	113	71	184
Conclusions	216	197	413
Total	1231	985	2216

The graduate level has 1231 elements and the undergraduate 985. It is relevant mentioning that some elements were not found in the corpus document. Each thesis is written by the students based on the institutional guidelines of each university, these academic guides are derived from books by authors of research methodology. However, most of the guides coincide with the definition of each element of a thesis. In our work, we have selected the most common thesis elements: problem statement, justification, objective, research questions, hypothesis, methodology and conclusions. Note that each experiment performed to analyze some characteristic of a thesis, required a specific sub-corpus, for instance, to the analysis of the connection between the objective and the conclusions section, only the objective and conclusions elements were used.

3.1 Subcorpus Used in Conceptual Sequence

The sub-corpus consists of 240 collected samples, 120 samples for graduate and 120 for undergraduate level, with 40 samples of each of the sections: problem statement, justification, and conclusion. The second kind included documents of Bachelor and Advanced College-level Technician level.

3.2 Subcorpus Used for Weak Sentences

Identifying a weak sentence in the conclusion is complicated for human annotators, since it requires expertise in computer science thesis advising and the ability to discern if a sentence in a conclusion complies with minimum requirements. The annotators were instructors from public universities, teaching courses and advising students on their final year about their theses. The background of the annotators was in computer science, specifically in information technologies. Annotators tagged each sentence as strong or weak. For sentences identified as weak, annotators provided the type of weakness.

Annotators were provided with a guide defining the qualities of a good conclusion and including examples of weak and strong sentences. The criteria used by annotators to identify the two kinds of sentences, based on university institutional guides and authors of methodology books were: a global response to the research question, compliance with each of the research objectives, the acceptance or rejection of the hypothesis, and the contrast between the fundamentals and results.

Fifty-Five Advanced College-level Technician degree (TSU) level theses were gathered. Then, from the conclusion section, 544 sentences were obtained. Finally, the sentences were sent to human annotators to be tagged with strong and weak classes. In addition to that, 210 Bachelor and Master theses were collected. These documents were used by the component IWS after doing Unsupervised Clustering. The aim of the clustering was to identify sentences that are representative of concepts found in approved theses. This clustering was used only in the component IWS [14].

3.3 Subcorpus Used in Justification Experiments

The total sentences extracted from the justification section of the corpus were 1006, that includes 275 sentences related to some elements and 731 sentences without any element identified. Table 2 details the number of sentences describing elements.

Table 2. Experimental collection

Essential element	Sentences	Train set	Test set
Importance	65	43	22
Necessity	55	36	19
Convenience	78	52	26
Benefits	77	51	26
Total	275		

Two annotators with academic experience marked sentences with the elements of interest, i.e. importance, necessity, convenience, and benefits. The paragraph sentences are part of each of the justifications of the collected corpus.

3.4 Human Reviewers

For validation, a set of elements from the corpus was selected for each experiment. In the following subsections, each experiment defines its own set of validation. The developed methods are supported by the Fleiss and Cohen Kappa Test. A group of human reviewers was employed for the annotation process.

Human reviewers oversaw annotating the validation sets. Thus, a reference set (gold standard) could be obtained. The reviewers are instructors (lecturers) from public universities in the computing field. They have expertise in reviewing theses. Finally, the collected corpus was made available to the community on the website: www.coltypi.org. The full corpus was published with an option to download elements of a given thesis.

4 Method to Analyze Conceptual Sequence

For this experiment, we selected the sections of *Problem Statement, Justification and Conclusions* because they are longer texts containing multiple sentences and this allows us to generate paragraph models. In contrast, an Objective or Hypothesis can hardly generate a model, since typically consists of a sentence. Below, we provide the general scheme to evaluate the Conceptual Sequence (Fig. 2).

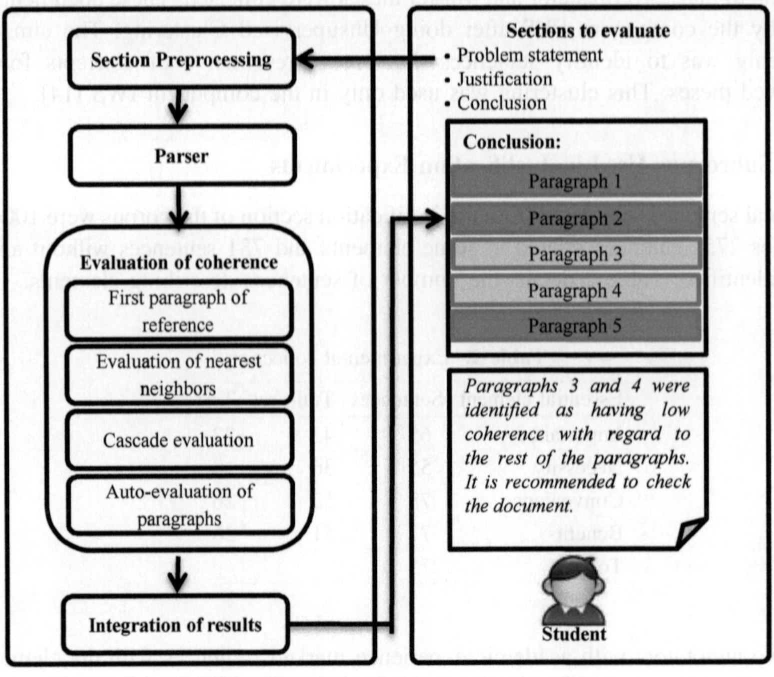

Fig. 2. Conceptual sequence scheme.

Our model incorporates four different schemes (methods) to evaluate the conceptual sequences and each is applied depending on the section that is being evaluated. For Problem Statement and Conclusions, First Paragraph of Reference (FPR) and Evaluation of Nearest Neighbor (ENN) methods are appropriate. In the case of the Justification section, Cascade Evaluation (CE), ENN, and Auto-evaluation of Paragraphs (AP) are pertinent.

Conceptual sequence analysis incorporates different schemes, and each is applied depending on the section that is being assessed. A document with a suitable structure presents a clear sequence of topics through their paragraphs. For example, in a conclusion section, the paragraphs are connected by the same main topic, and contains paragraphs that support the results, considering the topic of the problem statement.

First Paragraph of Reference (FPR): This method identifies the transitions that originate in the first paragraph of the conclusion, also in subsequent paragraphs. For instance, the term "system" may appear in a first sentence as subject, later the same term may appear as an object. Afterwards, in the last sentences the same term may appear as the object. These transitions are of interest because they indicate that concepts of the first paragraph are identified throughout the conclusion.

Evaluation of Nearest Neighbor (ENN): This is a scheme designed to evaluate null paragraphs identified after being examined with the FPR scheme. So, its main purpose is to relate null paragraphs with their prior or subsequent neighbor.

Cascade Evaluation (CE): It identifies the transitions that appear throughout the conclusion but in a distributed way, that is, the transitions are not concentrated in a single position.

Auto-evaluation of Paragraphs (AP): This was designed to evaluate null paragraphs that remained after being evaluated with any of the previous schemes. The definition of each of the models is detailed in [15].

The first step of the method is the preprocessing of the student writing. This step is done in two main tasks. The first focuses on segmenting each section of a draft into paragraphs. Each paragraph is defined as a sequence of sentences and is bound by the line feed. EGrid tool requires as input, the text in a Treebank format[1]. The second task follows the above procedure and is a process of translation from Spanish to English using Google Translator since our corpus is in Spanish. A comparative study of several translation systems such as Google, Word-Lingo, Systran, showed that Google translator showed the best performance [16].

The result of the translation enables processing the text with an English parser, in particular the Stanford parser. Currently, the parsers for Spanish do not adhere to the Treebank tags, as English parsers. Since we are mainly interested in the entities, we just expect that the translator keeps consistent translating such entities that in most cases occurred, given that they are mainly technical terms.

[1] https://www.sketchengine.co.uk/penn-treebank-tagset.

4.1 Experimental Results

The objective of the experiments was to apply the method and the previously designed evaluation schemes on the corpora. In this way, it was generated a diagnosis of both levels in the Problem Statements, Justifications and Conclusions. In the experiments, the method was applied using the different schemes included. The remaining 90% of the corpus was used in each section.

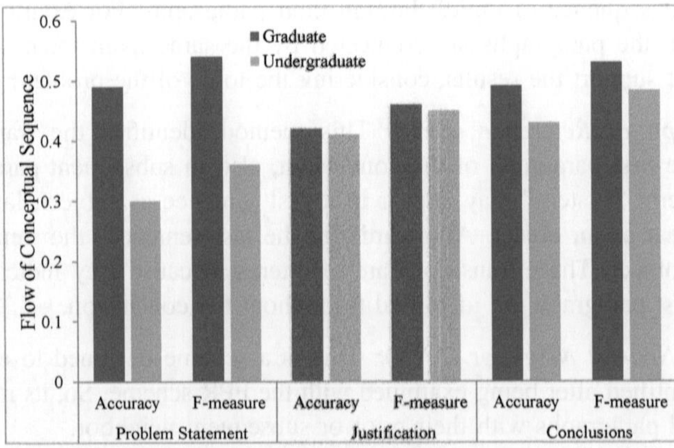

Fig. 3. Results of experimentation

Figure 3 summarizes the results in terms of average Accuracy and F-Measure values provided by the tool, for the three sections. These results show that grad students' paragraphs from the corpus are better linked than those of the undergrad students, being wider the difference in Problem Statements and closer in Justification. The results were qualitatively validated in the original text, i.e. it was looked for the relation of paragraphs identified by the schemes in the conclusions.

4.2 Evaluation of Schemes

Furthermore, a qualitative analysis was conducted on the examples of the corpus and it was observed that the schemes allow identifying topic changes within the section evaluated. For example, if a conclusion had a low value of connectivity, this meant that there was evidence that the conclusion contained several topics, instead if a conclusion showed high connectivity the conclusion was more homogeneous regarding the main topic.

Table 3. Results of evaluation

Section	Correspondence	No correspondence
Problem statement	0.64	0.36
Justification	0.70	0.30
Conclusion	0.83	0.17

Thus, it was decided to compare our schemes against a Bayesian segmentation approach, which was driven with a focus on lexical cohesion [17]. The segmentation task divides a text into a linear sequence of topically coherent segments. The authors argue that well-formed texts induce to consistent lexical distributions. The whole corpus was considered for evaluation (i.e. 240 samples). Each element of the corpus was evaluated using both techniques. If the methods identified the same topics that segmentation algorithm did, then the result was tagged as having a Correspondence; otherwise the result was No Correspondence. Table 3 summarizes the percentages of examples for each case, agreeing mostly in the three sections, with a higher agreement in Conclusions.

In [18] O'Rourke and Calvo evaluated the flow of paragraphs in university essays (English language) using two techniques MFN and SVD. The essays were classified into two groups, group A corresponds to essays with a score between 60 and 70 and group B with scores between 70 and 100. The authors found that the test group B obtained greater semantic flow than group A. However, the measure effect size, which determines the strength of the difference between two groups, was 0.18. When calculating the effect size for the methods developed in this thesis, a value of 0.59 among graduate and undergraduate levels was obtained. This result means that the methods can differentiate the two levels. Besides, it was a good result considering the result of the work described above.

4.3 Discussion

Assessing the flow of concepts in proposal drafts is a complex task for computers, and sometimes even for humans. It was understood that the behavior of transitions in the Conclusions adhere to a pattern where most of the central entities concentrated in the beginning of the Grid, i.e. the first paragraph contains information that was further developed in the other paragraphs. This behavior corresponds to a pattern, which begins with the restatement of the main premise, then a summarization of the key points, and finally the formulation of recommendations, assessments and forecasts, as expressed in some academic writing guidelines. The similarity between this pattern and that observed in the EGrid was verified by reviewing the text of the Conclusions of our corpus: it was observed that when the student redefined the main problem, he/she used many of the key terms of his proposal, which were reflected in the subsequent paragraphs.

The problem statement section showed a similar pattern as the Conclusions. However, a slight difference in some cases was that most of entities appeared in the first two paragraphs. Regarding the Justification section, it was observed that different entities were referred in a chained way, corresponding to different issues discussed at a time, as expected for this kind of section. We can assume that if a conclusion or the problem statement is like the patterns identified in the corpus, then that document is likely to have good writing. Finally, the method with their different schemes can be easily applied directly to English drafts, only by omitting the translation step that was needed for Spanish. Moreover, student drafts in other domains can also be analyzed without too much trouble.

5 Method to Analyze Weak Sentences

The method has a Weak Sentences Identifier module, which contains three main components. The first component Identifying Weak Sentences (IWS) is responsible for discerning whether a sentence is weak or strong. It includes five models that use different techniques and resources, such as: lexical richness, similarity between sentences measurement, use of speculative terms and overlap with terms from the conclusion sections of approved theses.

The second component Classifying Weak Sentences (CWS) looks at sentences that were identified as weak by the previous component and determines the kind of weakness. This component takes advantage of a corpus tagged by human annotators to train a model with the main weaknesses identified by them. Finally, the third component Customized Feedback to Student and send feedback to student regarding the weakness type identified by the component CWS. Below, in Fig. 4 we show the scheme of Weak Sentences Identifier.

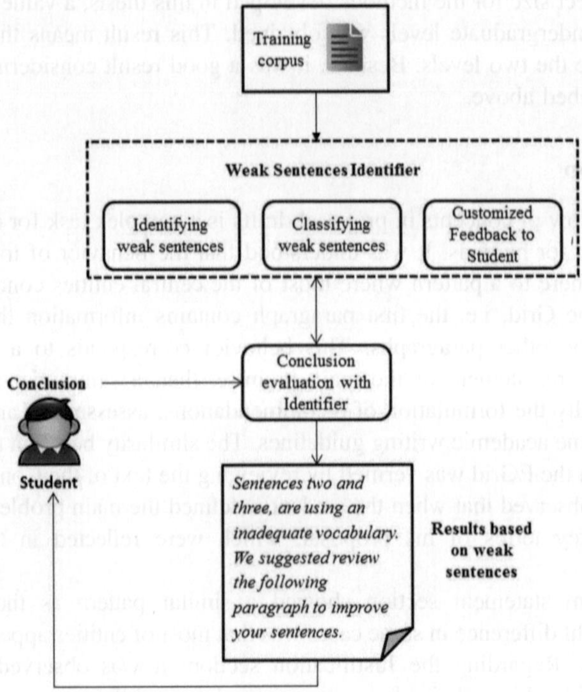

Fig. 4. Weak Sentences Identifier scheme.

The models developed seek to capture features that reflect weak or strong sentences. A sentence presents weakness when concepts are written in general instead of specific terms, or if there is absence of reflections and value judgments. These types of sentences do not fit in the conclusion section. An example of a weak sentence, part of a thesis about computer networks, states:

Security should not be a problem, neither for networks nor in everyday life, but as some humans do not have a social conscience either by greed, a bad curiosity, ambition.

One can notice in the sentence that the student is using speculative words such as "should". Also, the student expresses a philosophic argument that fits better into an introduction section as a motivation of the problem.

A strong sentence means that the text is reasonable and makes sense in a conclusion. For instance:

The new system will help to lower costs for man/hours invested in the maintenance of the infrastructure.

We can see in the sentence that the student is writing a possible consequence of the implementation of the system. The components of Weak Sentences Identifier method are detailed in [12].

5.1 Experimental Results

The first step was submitting the sentences of the conclusions to human annotators, with the aim of generating a gold standard. The sentences were tagged with the class "weak" and "strong", and for each "weak" sentence, a description of why it was considered "weak" was provided. A total of 168 sentences were tagged with weak class and 376 sentences with the strong class. Table 4 shows the confusion matrix of agreement and disagreement between annotators. Note, that most agreements were on strong sentences.

Table 4. Confusion matrix between annotators

Class	Weak	Strong
Weak	48	120
Strong	36	340

The Kappa-Cohen agreement between human annotators was of 0.25, corresponding to a Fair level of agreement. A third annotator adjudicated the disagreements between the two primary annotators.

5.2 Identifying Weak Sentences

Each sentence was processed with the models to obtain features. The results were used as input to a classifier (NaiveBayesMultinomial). Lexical Richness alone was used as baseline. Below, the results obtained by classifiers with 10-fold cross-validation are presented. Since the goal is to remedy "weak" sentences, the main interest is in the precision and recall of the "weak" class, but it is also in showing the performance of the "strong" class.

Table 5 shows that all models outperformed the baseline F-measure of 0.527. It was added to the baseline different model features with the goal of improving precision and recall. The system with the highest F-Measure for Weak sentences (0.622) was SC +SWE+CM2+LR, though SC+CM+LR achieved slightly higher precision (0.613 vs. 0.603). These results may suggest that identifying speculative words generally improves recall, though at a small cost of precision.

Table 5. Classifying results

Models	Precision	Recall	F-Measure	Class
LR*	0.556	0.5	0.527	Weak
	0.582	0.636	0.608	Strong
SC+SW+CM+LR	0.567	0.506	0.535	Weak
	0.589	0.647	0.617	Strong
SC+CM+LR	0.613	0.548	0.579	Weak
	0.624	0.685	0.653	Strong
SC+SW2+CM2+LR	0.603	0.643	**0.622**	Weak
	0.653	0.614	**0.633**	Strong

*LR: Lexical Richness; SC: Similarity of Cosine Words; SW: Speculative Words; SWE: Speculative Words Expanded; CM: Coverage Model

For the task, the purpose was to find a balance between precision and recall, like that of the SC+SWE+CM2+LR model, since the system must clearly distinguish a weak sentence in the conclusion -otherwise the system could confuse the student- and at the same time should have good coverage.

Bethard et al. [10] developed an intelligent tutor to identify science concepts and student misconceptions (Elementary level). One of the methods developed by the researchers was the identification of student misconceptions, which obtained a MAP of 64%. The developed IWS method reached an F- measure of 0.622.

5.3 Classifying the Weak Sentences

The models were evaluated to identify the kind of weakness using as gold standard the "weak" class. Those weaknesses were tagged with the types already identified from annotators' descriptions: NC[2], GT[3] and VO[4]. There were 113 examples of GT, 34 of NC and 18 of VO. Because of the small number of sentences with types NC and VO, they were merged into a single NC_VO class. The results (Table 6) were obtained by classifiers using 10-fold cross-validation.

In this experiment, the best combination of features was SW+SWE+CM+CM2+ Variety. Other combinations were tested that included density and sophistication (from the LR features) and cosine similarity models, but these did not perform as well. Sentence similarity features likely contribute less to this task because they are not focused on identifying vocabulary and term-based issues.

[2] NC: The sentence was not connected to research results.

[3] GT: The sentence is written in General Terms.

[4] VO: The vocabulary used in the sentence is not adequate.

Table 6. Classification results

Models	Precision	Recall	F-measure	Class
SW+SWE+CM+CM2+Variety	0.6	0.35	0.439	NC_VO
	0.748	0.894	0.815	GT

5.4 Discussion

The result of F-measure achieved by the method IWS was close to the method developed in [10]. However, the level of corpus used in this task could represent more complexity to IWS method identifying a weak sentence.

Also, it was found that weak and strong sentences share features, i.e. a weak sentence contained similar terms as a strong sentence. For instance, a weak sentence can be written in a descriptive manner and adjust to another section as introduction or justification. These differences allowed the classifiers to identify patterns.

Moreover, it was identified that the weak and strong sentences, labeled by the annotators, are not affected by the writing styles of each student. For instance, the use of the active voice or passive voice in a conclusion does not affect the tagging process of sentences, since the criteria defined to label weak or strong sentences contemplated own aspects of a conclusion, such as contrasting results. The amount of training data for low frequency weakness types needs to be increase, i.e. inadequate vocabulary. This would allow the method to have better coverage of the distinct types of weaknesses, and to strengthen the features of weak and strong sentences.

6 Method to Analyze Essential Elements in Justification

The developed method seeks to capture the essential elements (core dimensions) within the paragraphs of the justification section. To perform the analysis, we employed a language model approach. The first step was to create a model for each element from the training set, then to each sentence of test set was calculated its perplexity value, according to the corresponding language model. Two language models were trained: SRILM[5] (based on n-grams) and RNNLM[6] based on the implementation of recurrent neural networks by Mikolov [19].

The SRILM tool settings were taken as follows: 6-grams were used for all components, with smoothing employing the Ristad's natural method. The RNNLM settings of the neural network were: No classes were used, the back-tracking algorithm (6 times), and a hidden layer of 50 neurons was defined.

Text Preprocessing: Punctuation symbols and citations were removed, and numbers were replaced with the token "NUM". Then, the text was lemmatized with Freeling (http://nlp.lsi.upc.edu/freeling/) with the default config file for Spanish, that also allowed to obtain the part-of-speech (POS) tags for the text.

---

5 www.speech.sri.com/projects/srilm/.

6 http://www.fit.vutbr.cz/~imikolov/rnnlm/.

Data Preparation: Each section was tested separately. For experiments, two subsets were obtained, one with 67% for training and the other with 33% for testing. With the training set, the model was built, for each of the essential elements (one with RNNLM and one with SRILM) with and without POS tags. Thereafter, each language model was applied to the test sets of all dimensions. Two hundred and fifty-one sentences were randomly selected out of the 731 sentences without identified element and added for testing purposes as negative samples.

The output of each language model tool is numerical, which represents perplexity that expresses the confidence of the sentence tested in the model. A low value of perplexity indicates higher confidence. The measure used to compare the similarity was "average perplexity per word".

To interpret the results of perplexity, the F-measure was calculated, and the contingency matrix was created for each element. To build the matrix, we considered the annotated sentences by reviewer, that is, if a sentence obtained a low value of perplexity, this sentence was considered as a positive element (true positive). However, if the test sentence obtains a high value of perplexity means that sentence was not labeled in any of the categories (true negative) by the annotators. In addition, the threshold used to determine whether a sentence was positive or negative was estimated by averaging the perplexity value obtained by annotators, for each of the elements.

6.1 Experimental Results

First, we processed each of sentences of the test set in each of the four-language models already built. Below in Table 7, we show the result in terms of weighted F-measure (*WFm*), which considers the number of elements per class, and is used to counteract the unbalanced classes, as expressed in:

$$WFm = FM_p * \left(\frac{P}{P+N}\right) + FM_n * \left(\frac{N}{P+N}\right) \tag{1}$$

where *FMp* and *FMn* are the F-measures for the positive and negative classes, and P and N are the number of positive and negative instances, respectively.

We can notice in Table 7 that the language model based on n-grams (SRILM) achieves better values than the second model, either with or without POS tags, identifying the elements of interest. Part of speech tagging provides additional information that allows capturing more relations in the language model, causing a noticeable improvement in both kinds of language models.

Table 7. Essential element results (WFm)

Evaluated element	Language model methods			
	SRILM	SRILM+POS	RNNLM	RNNLM+POS
Importance	0.913	**0.919**	0.883	0.910
Necessity	0.895	**0.925**	0.886	0.918
Convenience	**0.898**	**0.898**	0.882	0.892
Benefits	0.879	**0.902**	0.821	0.878

Regarding the elements of the justification, Necessity was more clearly identified by the language model than the rest, despite having less instances. However, both models also show a good discrimination capability for Importance. Benefits also reached acceptable value by SRILM but lower with RNNLM. Convenience, with a higher number of instances, seems harder to identify, according to the results produced by both language model approaches.

6.2 Discussion

The size of the experimental collection is relatively small, impacting on the trained language models. We have a plan to extend the corpus. We are also considering word representations [20] that promise to improve the efficacy of the models. It was expected that the language model based on recurrent neural networks have better performance, but the short length of each sentence may have caused that this did not happen. N-grams seem to work better on this task.

Besides POS information, we are exploring other features to improve the discriminative power of models. Once we reach an acceptable classification level for the different elements, we will incorporate the language models in an online application to evaluate student drafts, providing feedback when needed.

7 Conclusions

Communicating ideas, through written language, is essential to the knowledge society. Many efforts to improve the students' writing at early stages of their education have been proposed. Research studies dealing with the use of technology as an aid for writing have grown. In the study of this problem, considerable efforts were identified aiming to guide students towards better performances; for example, intelligent tutoring and platforms managing the progress of students by sending personalized feedback. Moreover, automatic systems were detected whose purpose is to evaluate distinct aspects such as lexical richness, coherence and cohesion; mainly in students' essays. In addition, a smaller amount of studies related to the internal connection of essays was pinpointed.

NLP techniques and resources helped to solve specific problems of this analysis, however some techniques were explored in depth to adapt and achieve objectives. For some problems, the design of a specific methodology was required as the analysis of flow of concepts or the identification of weak sentences in a conclusion section.

In this chapter, we have presented the analyses of large sections of a thesis: problem statement, justifications and conclusion, seeking to cover the analysis of the elements of a thesis. During the experiment, we have identified that graduate students obtain better evaluations. Despite the methods were designed to bring support mainly to undergraduate students, graduate students could also use the methods designed in their draft theses. We are currently developing an online system available at www.turet.com.mx[7],

[7] Click on the image "TURET 2.0".

so that students have access to the evaluation of the thesis. Even though the methods were designed for Spanish language documents, these methods can be addressed to other problems and they could be used in English language works. For instance, it is possible to address other problems such as the identification of weak sentences in essays written in English by L2 students. It would however be necessary to tag a set of sentences to detect the main errors in the essays.

Future work is expected to embed these methods in intelligent tutors to conduct a pilot test with students from different universities. In addition, we will try to perform the analysis with all elements of the corpus, as Coltypi has been growing, as well as seeking to evaluate other sections of a thesis such as the section of justification and methodology.

Another future work will be the evaluation of the content, together with the assessment of the structure as developed in this chapter. For example, between the objective and conclusions sections, a coverage model was explored in this work, however, a challenge to further work would be to assess the semantic level of coverage between those two sections.

References

1. Bitchener, J., Basturkmen, H.: Perceptions of the difficulties of postgraduate L2 thesis students writing the discussion section. J. Engl. Acad. Purp. **5**, 4–18 (2006)
2. Allen, G.: The Graduate Students' Guide to Theses and Dissertations: A Practical Manual for Writing and Research. Jossey-Bass Inc Pub, San Francisco (1976)
3. Davis, J., Liss, R.: Effective Academic Writing 3, The Essay. Oxford University Press, Oxford (2006)
4. Webber, B., Egg, M., Kordoni, V.: Discourse structure and language technology. Nat. Lang. Eng. **18**, 437–490 (2011)
5. O'Rourke, S., Calvo, R.: Analysing semantic flow in academic writing. In: Proceedings Conference on Artificial Intelligence in Education: Building Learning Systems that Care, pp. 173–180. IOS Press, Amsterdam (2009)
6. Barzilay, R., Lapata, M.: Modeling local coherence: an entity-based approach. Comput. Linguist. **34**, 1–34 (2008)
7. Crossley, S., Roscoe, R., McNamara, D.: Using automatic scoring models to detect changes in student writing in an intelligent tutoring system. In: Proceedings of the 26th FLAIRS, pp. 208–213 (2013)
8. Olney, A.M., et al.: Guru: a computer tutor that models expert human tutors. In: Cerri, S.A., Clancey, W.J., Papadourakis, G., Panourgia, K. (eds.) ITS 2012. LNCS, vol. 7315, pp. 256–261. Springer, Heidelberg (2012). https://doi.org/10.1007/978-3-642-30950-2_32
9. Graesser, A., et al.: The relationship between affective states and dialog patterns during interactions with AutoTutor. Interact. Learn. Res. **19**, 293–312 (2008)
10. Bethard, S., Okoye, I., Sultan, A., Hang, H., Martin, J., Sumner, T.: Identifying science concepts and student misconceptions in an interactive essay writing tutor. In: Proceedings of the 7th Workshop on Building Educational Applications Using NLP, pp. 12–21 (2012)
11. Burstein, J., Marcu, D.: A machine learning approach for identification of thesis and conclusion statements in student essays. Comput. Humanit. **37**, 455–467 (2003)

12. López, S.G., Bethard, S., López-López, A.: Identifying weak sentences in student drafts: a tutoring system. In: Mascio, T.D., Gennari, R., Vitorini, P., Vicari, R., de la Prieta, F. (eds.) Methodologies and Intelligent Systems for Technology Enhanced Learning. AISC, vol. 292, pp. 77–85. Springer, Cham (2014). https://doi.org/10.1007/978-3-319-07698-0_10
13. Daudaravicius, V.: Automated evaluation of scientific writing: AESW shared task proposal. In: Proceedings of the Tenth Workshop on Innovative Use of NLP for Building Educational Applications, pp. 56–63. Association for Computational Linguistics (2015)
14. Bellegarda, J.: Unsupervised document clustering using multiresolution latent semantic density analysis. In: Workshop on Machine Learning for Signal Processing, pp. 361–366 (2010)
15. González-López, S., López-López, A.: Analysis of concept sequencing in student drafts. In: Rensing, C., de Freitas, S., Ley, T., Muñoz-Merino, P.J. (eds.) EC-TEL 2014. LNCS, vol. 8719, pp. 422–427. Springer, Cham (2014). https://doi.org/10.1007/978-3-319-11200-8_36
16. Aiken, M., Ghosh, K., Wee, J., Vanjani, M.: An evaluation of the accuracy of online translation systems. Commun. IIMA **9**(4), 67–84 (2009)
17. Eisenstein, J., Barzilay, R.: Bayesian unsupervised topic segmentation. In: Proceedings of the Conference on Empirical Methods in Natural Language Processing, pp. 334–343 (2008)
18. O'Rourke, S., Calvo, R.: Analysing semantic flow in academic writing. In: Proceedings of the 2009 Conference on Artificial Intelligence in Education: Building Learning Systems that Care, pp. 173–180 (2009)
19. Mikolov, T., Deoras, A., Kombrink, S., Burget, L., Cernocký, J.: Empirical evaluation and combination of advanced language modeling techniques. In: INTERSPEECH, pp. 605–608 (2011)
20. Turian, J., Ratinov, L., Bengio, Y.: Word representations: a simple and general method for semi-supervised learning. In: Proceedings of the 48th Annual Meeting of the Association for Computational Linguistics, pp. 384–394. Association for Computational Linguistics (2010)

Helping MOOC Teachers Do Their Job

Ranilson Paiva(✉) and Ig Ibert Bittencourt

Federal University of Alagoas (UFAL) - Computing Institute (IC),
Av. Lourival Melo Mota, s/n - Tabuleiro do Martins, Maceió, AL 57072-970, Brazil
ranilsonpaiva@ic.ufal.br, ig.ibert@gmail.com

Abstract. There is a quest to provide education from anywhere, at any time and for anyone, using digital information and communication technologies. However, there is no equivalent increase in support for the instructors responsible for maintaining such courses, which is evidenced by the large number of dropouts and failures in such courses, to which learners justify as a lack of support from instructors. Instructors, however, complaint about the huge effort to manage such courses. In order to provide this support, instructors would have to: (1) to discover situations of pedagogical interest occurring in their courses; (2) understand these situations; (3) make decisions to address them; (4) monitor and evaluate the impact of the decision made. However, instructors do not master these abilities, nor is it practical or appropriate to ask them to do so. We propose a process, and an authoring solution that implements it, to guide pedagogical decision-making in online learning environments. Our proposal is based on decision-making informed by educational data and data visualization, with the assistance of an authoring system to promote cooperation between artificial intelligence and human intelligence. We conducted experiments, using a MOOC named MeuTutor-ENEM, to evaluate the process and the authoring solution, and the results indicate that the process helped instructors make better pedagogical decisions, and the authoring solution was positively perceived by the instructors.

1 Introduction

We are experiencing changes in the educational paradigm, where digital information and communication technologies is used to provide education from anywhere, at any time and for anyone (*AAA Learning*) [4]. This fact is evidenced by yearly increases in global interest in distance learning [7,9]. Among the types of distance courses, MOOCs[1] are gaining attention due to their potential to "propagate and democratize education" [27].

However, the dynamics of a MOOC bring challenges to the instructors responsible for maintaining it. For instance, in a course titled *Bioelectricity: A Quantitative Approach*, only 2,46% of the learners we able to finish the course (earn a certificate) [2]. That was not an isolated case. Researchers[2] [16] estimate that,

[1] MOOCs: Massive Open Online Courses.

[2] Check: http://www.katyjordan.com/MOOCproject.html.

A. I. Cristea et al. (Eds.): HEFA 2017, CCIS 832, pp. 52–67, 2018.
https://doi.org/10.1007/978-3-319-97934-2_4

on average, 85% of learners dropout from MOOCs, justifying it as **a lack of support by their instructors** (teachers and/or tutors) [19].

In order to provide this support, instructors would need to analyze data from the learners' interactions in search for educational profiles and relevant information for the teaching and learning process [24], in order to support pedagogical decision-making [3]. However, instructors do not receive appropriate technology aid to deal with these data [26], giving rise to barriers against technology integration and instructional practice [15].

These facts point to the need to direct and aid these instructors in a way machine intelligence complements human intelligence [1,6,25]. Based on that, we asked ourselves the following question: *How can we use digital information and communication technologies to guide instructors' pedagogical decision-making?*

In order to address this question, we defined the following research objectives:

1. RO1: Design a process that systematically guides pedagogical decision making;
2. RO2: Create an authoring solution, based on the process, to build informed pedagogical decisions, in course time, for online learning environments;
3. RO3: Enable the authoring of pedagogical decisions at different levels of granularity without affecting the users' perception about the authoring tool;
4. RO4: Define visual metaphors to allow teachers and tutors, transparently, to use techniques of educational data processing and analysis, to inform them of relevant pedagogical situations occurring during course time.

As a solution proposal, we created a process (the Pedagogical Decision-Making Process - PDMP) to guide pedagogical decision-making for instructors of online learning environments (OLEs), and an authoring solution (T-Partner) that implements this process. Initially, we manually applied the process in an online course [22], with the intention of helping instructors use the educational data generated by the studied learning environment to: (1) discover situations of pedagogical interest occurring in their courses; (2) understand these situations; (3) make decisions to address them; (4) monitor and evaluate the impact of the decisions made. These were the 4 steps of the PDMP.

The authoring solution (T-Parner, in a proof-of-concept stage) implements the process, so that we could allow instructors to use educational data analysis techniques (without the need to learn them), in order to detect and address pedagogical situations occurring within a course, such as: detecting performance problems within a group, comparing groups, predicting dropouts and predicting inadequate performance in a group. Our proposal is based on informed decision-making by educational data, promoting cooperation between artificial intelligence and human intelligence. This means we use artificial intelligence to augment human intelligence.

In order to test our proposal, we conducted two experiments using the educational data from a MOOC (MeuTutor-ENEM). In the first, we manually applied the PDMP to help teachers improve their learners' interactions. In the second, we developed the interfaces of the T-Partner. It was possible to interact with its functionalities, but most of them were not implemented. For that, we invited professors, teachers and tutors (the professionals for which we intend to create the solution) to interact with these interfaces. The objective was to evaluate if the tool assisted the instructors in making decisions to address a pedagogical situation. The results for the first experiment was an increase (more than 10%) in the learners' interactions. For the second experiment, participants showed positive perceptions about T-Partner's interfaces[3].

This chapter is organized as follows: in Sect. 3, we present the main ideas about our proposal. In Sect. 4, we describe the two experiments we ran to evaluate our proposal (one for the process and the other for the authoring solution). In Sect. 5, we summarize the results of the experiments and, finally, in Sect. 6, we draw our conclusions, based on the results of our experiments.

2 Related Work

There is a vast number of authoring tools and authoring models [10,17,18], most of them are focused on pre-instruction (assisting the creation of the adaptive learning environment and the adaptive content) [10], an example is the MOT (My Online Teacher) adaptation authoring tool set that includes MOT3.0, a content authoring and labeling tool, and the PEAL, which objective is to author adaptation strategies [11]. Another example is the ALAT, which objective is to allow non-technical authors to easily specify the desired adaptation to improve user satisfaction, learning etc [8].

The authoring solution presented in this chapter differs from those other. T-partner assists authors in the instruction phase (from the beginning of the course until its end). We performed a systematic literature review (SLR) [10] to search for related work. We looked for authoring tools that: (1) were focused on assisting the pedagogical activities of teachers/tutors; (2) supported educational decision-making; (3) acted in the instruction phase, that is, authoring happens during the time of the course; (4) were supported by educational data; (5) acted in specific groups of students. These are the characteristics of the authoring tool proposed in this chapter.

EDUCA 2.0 is an Authoring Tool for the creation of adaptive learning material, which is made available to students through mobile devices [5]. EDUCA uses Self-Organizing Maps (SOMs), to identify the learning style of each student and thus adapt the content accordingly. EDUCA presents the ability to make personalized recommendations based on the learning characteristics of the students. However, such a decision is taken without the participation of the teacher whose activity is restricted to creating content.

ProTracer 2.0 is an educational system (Intelligent Tutor System) that promotes programming teaching by tracking students' actions to identify problems and provide specific help to learners. The system can also help teachers evaluate

[3] Due to limitations in the number of pages for this chapter, we had to focus on the essentials. This way, the description of the proposal, the description of the experiments and the results were summarized to their core.

the progress of their students through a virtual assistant [6]. However, the system is used in a very restricted domain. Apparently, the teacher's function is to aggregate information to the system and to create content (tips, new activities and new lessons).

ASSISTments is a platform that allows teachers to create, or use pre-built, artifacts called ASSISTments, group them in a set of problems and assign them to their students. The system gives immediate feedback to students while they are studying and provides student-level data to teachers on any task. The platform allows any user, primarily researchers, to create controlled and randomized experiments on content [13]. Basically, the teacher can only create questions and follow the students' results to the questions and, if necessary, create new questions or improve those already created. The ASSISTments allow the evaluation of decisions, but this resource is used more commonly by trained researchers and not by teachers/tutors.

WISE (Web-based Inquiry Science Environment) is a free, open source curriculum platform that provides users with instructional design support through evidence-based refinement of learning material. Users are teachers, tutors, curriculum developers, etc. [29]. However, WISE does not aim to solve specific pedagogical situations being experienced by the students during a course, but rather refine learning material.

3 Proposal

In this section, we briefly present the Pedagogical Decision-Making Process (PDMP) and the authoring solution that implements it (T-Partner).

3.1 Pedagogical Decision-Making Process (PDMP)

The process (Fig. 1) was created [21–23] to guide online learning environment instructors to: (1) to discover situations of pedagogical interest occurring in their courses; (2) understand these situations; (3) make decisions to address them; (4) monitor and evaluate the impact of the decisions made.

The PDMP[4] is a cyclical process and consists of two phases: the construction phase and the implementation phase. In the construction phase there is the collaborative interaction of human and computational intelligences, to create computational artifacts called *Pedagogical Decision Capsule*. These artifacts are the encapsulation of the instructors' decisions about which pedagogical situation they want check in a learning environment, what decisions to take to address such situation, and how they wish to assess whether the decisions have been effective or not. The purpose is to get the best of each intelligence involved (human and computational), that is, human intelligence is called for the decision-making capacity, but for this computational intelligence processes the educational data, extracting information from them to support those decisions.

[4] More information about the PDMP can be found in [22].

Fig. 1. The pedagogical decision-making process.

The PDMP was used in some researches to evaluate the effectiveness of gamification elements in educational environments [20], to measure the differences of male and female students' interactions in an OLE [28], to improve students' interactions in an OLE [23] among other uses.

3.2 T-Partner

Following the process (PDMP) manually is not an easy task and is subject to failures (human). We need a technological solution to support instructors, the T-Partner[5] to author *Pedagogical Decision Capsules.*

T-Partner will need to be integrated to the learning environment in order to access data about: learners, educational resources available, pedagogical approaches offered, other educational data resulting from interactions (user-user, user-content and user-environment), among other data that can be used to support pedagogical decision making.

In Fig. 2, we present an overview of the integration between T-Partner and an online learning environment: (1) learners interact with the online learning environment (OLE); (2) these interactions generate educational data that are stored in the OLE's database; (3) these data are retrieved and processed by the T-Partner; (4) the processing results are used to inform instructors about pedagogical situations occurring in the OLE; (5) instructors use this information to make pedagogical decisions; (6) these decisions use the educational resources from the OLE; (7) the decisions should consider the OLE's interface capabilities; (8) the decisions are sent to the specific learners.

The T-Partner was, initially, idealized in two versions:

1. ***Light Weight***: it is an easy to use version, but it is more limited. This version is intended for users with little experience with computers and for the

[5] Teachers' Partner.

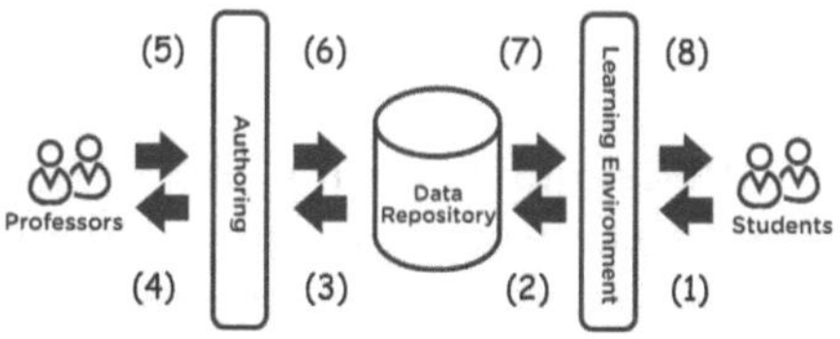

Fig. 2. Overall representation of the authoring solution behavior.

creation of pedagogical decision capsules with simple content or in a faster way. It can also be used as an entry version for beginners.
2. ***Heavy Weight***: it is a version with more features, but with higher complexity. This version is intended to allow greater detailing of the pedagogical decision capsules, which is appropriate if the user wishes to have more control, and requires a more detailed and specific use of the system.

Both versions of the authoring solution assist instructors: (1) search educational data about a pedagogical issue; (2) process this data; (3) generate visualizations based on the data processing; (4) recommend pedagogical approaches to deal with the pedagogical issue; (5) monitor and evaluate the decisions made; and (6) encapsulates the successful recommendations. However, in the heavyweight version, instructors need to explicitly define how they want the authoring tool to perform these tasks. For example, for task 2 (process educational data), instructors need to choose (among the available options) how they want to pre-process (filter, transform or remove), process (choose an available algorithm) and post-process (choose how to present the results) the data.

3.3 Implementing the PDMP

In this section, we describe how the T-Partner implements the Pedagogical Decision-Making Process (PDMP).

As mentioned in Sect. 3.1, the PDMP contains two phases. In the first, called *construction phase*, instructors act in parallel with the system to define the "hypothesis" they want to check in the educational data, investigate evidences about the"hypothesis", decide on what to do based on these evidences and check whether their decisions were effective or not, thus following the four steps in this phase. In the second, called, *implementation phase*, the successful decisions (those that were effective) are encapsulated and executed, automatically, by the system (check Fig. 1). Below, we better describe the four steps from the construction phase:

1. **Step 1: Define the Pedagogical Situation:** In this step, instructors should choose and personalize the pedagogical action of interest (among those available) and other details pertinent to each action. Instructors then define how the results should be classified, that is, what values characterize a result as inappropriate, insufficient or adequate (level of learning). The system searches

the data requested by the teacher, based on the chosen pedagogical action, and classifies it.
2. **Step 2: Investigate Pedagogical Situation:** In this step, the instructor selects the filters that he/she wants to apply in the data, as well as how he/she wishes to visualize the results of the processing. The system processes the data using an algorithm associated with the chosen pedagogical action and generates the visualization of the result. This step should provide information for pedagogical decision-making.
3. **Step 3: Define Pedagogic Decision:** In this step, the instructor defines the tasks (among those available in the OLE) for the learners, according to the levels of learning (inadequate, insufficient or appropriate).
4. **Step 4: Define Assessment:** In this step, the instructor should set the desired percentage for adherence (percentage of learners who followed a recommendation) and the outcome that he expects to obtain from those who followed them, for each level of learning.

4 Experimentation

In this section, we briefly present the experiments we did to evaluate the process and the authoring solution.

4.1 Experiment 1

We called this experiment: Evaluation of the Pedagogical Decision-Making Process. Its objective was to evaluate whether the process was effective, or not, in helping instructors make better decision using educational data, in order to meet our first research goal (check Sect. 1). In it, teachers manually applied the PDMP on a live MOOC (live), considering the educational data from 36 learners who were (randomly) selected to participate. The data analysis part was made by specialists involved in the experiment. Their task was to translate the teachers' needs (mimicking an automatic computer analysis). Below, we describe the steps followed in this experiment:

- **Step 1 - Define a Pedagogical Situation:** Collected data from 36 learners who were active for at least 6 months. Teachers defined the pedagogical situation of interest was as the evaluation of the students' interactions within the MOOC.
- **Step 2 - Investigate a Pedagogical Scenario:** We processed the educational data and classified the learners' interaction in: individual learning (those that directly influences learning, like: watching videos, solving problems, etc.), collaboration (those that helps making the MOOC better or helps other learner, like: reporting errors, rating videos, etc.), gamification (those focused on the game elements available, like: earning badges, completing missions, etc.) and social (those focused on interacting with other learners without the need of pedagogical gains, like: chatting, making friends, etc.). The results, for each learner, were plotted on a spider web graph and presented to the instructors for their consideration.

- **Step 3 - Define Recommendations:** After studying the visualizations, the instructors were informed about the educational resources[6] available in the MOOC and were told to decide what to do for each situation. They decided to recommend learners to interact more with the resources associated to the two classes (individual learning, collaboration, gamification and social) they had the least number of interactions. The learners had 30 days to complete the tasks recommended.
- **Step 4 - Define Success Criteria:** As a success criteria, instructors defined an adherence percentage[7] of 73%[8] and an increase, of at least 10%, in the learners' interactions with the recommended classes.

4.2 Experiment 2

In this experiment, we invited instructors (professors, teachers and tutors) to evaluate the T-Partner, regarding the following metrics: (1) utility perception - UP; (2) ease of use perception - EUP; (3) attitude towards use - ATU; (4) intention to use - IU; (5) correctness on the tasks requested - CTR; (6) perception about the Traffic Light Metaphor - TLM; and (7) perception about the terms that categorize student learning - TSL. Metrics 1 to 4 were based on the works of Teo et al. [30–32]. Participants were randomly redirected to one of three possible versions of T-Partner: (1) lightweight; (2) heavyweight; (3) manual (spreadsheet).

The lightweight (LW) and heavyweight (HW) are versions with different granularities of the proposed authoring solution, that is, these versions explore different combinations between artificial intelligence and human intelligence. The manual version (*spreadsheet* - SS) is a control version that simulates authorship, following the PDMP, in a very simplified way, using a familiar resource (spreadsheets). The objective was to evaluate the perception of the versions of the T-Partner (unknown) compared to the control version (familiar).

Each version covered the 4 steps of the PDMP (see Sect. 3.1). At each step, a participant should perform a task to construct a part of the pedagogical decision capsule and then respond to a questionnaire, expressing his/her perception about (1) utility; (2) ease of use; (3) attitude towards use; (4) intention to use.

From December, 12th, 2016 January, 1st, 2017, 244 participants interacted with the interfaces. Responses to the tasks (at each stage) and questionnaire responses were recorded for each screen and stored in a database. Data processing involved the removal of test records (39 records), records with anomalous data (1 record) and records of participants who were not professors, teachers nor tutors (29 records), resulting in 175 records. Finally, we performed the appropriate data analysis, according to [14].

[6] Educational resources are resources for teaching and learning, including complete courses, teaching material, modules, textbooks, videos, quizzes, educational software and any other tools or techniques used to support access to knowledge [12].

[7] The percentage of students that followed the instructors' recommendation.

[8] At least 26 learners followed the recommendation.

5 Results and Discussion

In the following sections, we summarize the results of the experiments described in the previous section, based on their respective research objective.

5.1 RO1: Design a Process that Systematically Guides Pedagogical Decision Making

In this section, we present the main results for experiment 1, that aimed at evaluating whether the process was effective, or not, in helping instructors make better decision using educational data.

During the six-month period learners freely interacted with the MOOC, we noticed that the 36 (randomly) selected learners made, on average, 0.25 friends per month; Rated, on average, 0.65 videos per month; Correctly answered, on average, 33.03 problems solved per month, and 39.80 problems solved incorrectly on average. These results were obtained dividing the absolute number of interactions (Fig. 3) by the number of months of interaction (6) and the result was divided by the number of learners selected (36). According to the teachers' evaluation, these learners' interactions were below the expected.

The learners received the personalized recommendations (based on their interactions), and had 30 days to complete these recommendations. After this period, their data were collected, this time with the following values: 0.30 friends; 0.70 videos rated; 30.96 correctly solved problems and 35.02 problems solved incorrectly, as shown in Fig. 3.

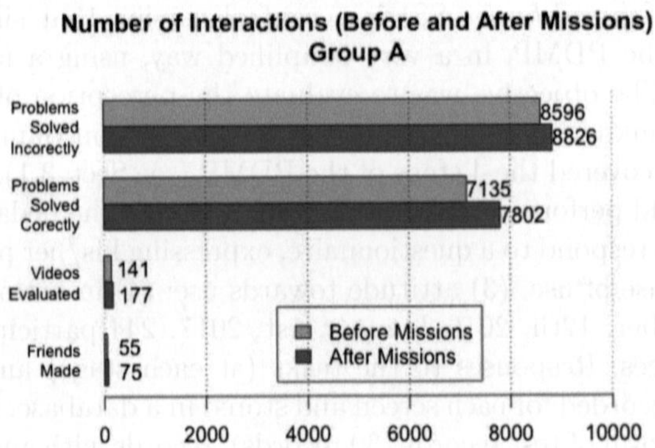

Fig. 3. Interactions before and after receiving the personalized recommendations (Missions).

In order to check if the learners' interactions before receiving the personalized recommendations[9] were different (statistical significance of 0.05) from their interaction after receiving the personalized recommendations, we ran the Wilcoxon

[9] Named in the learning environment as "Missions".

signed-rank test (data from a non-parametrical population and two measures from the same sample), as in Table 1.

Table 1. P-values for the Wilcoxon signed-rank test.

	Friends	Videos	Correct problems	Incorrect problems
Wilcoxon Signed-rank	0.001985	0.0006552	0.0002137	0.001656

These results show that the difference in the learners' interactions[10] before and after receiving the personalized recommendations is significant (statistically). As there were more interactions after the personalized recommendations, this suggests the recommendations created following the PDMP helped teachers improve students interactions when compared with the recommendations generated by the learning environment. The PDMP is a mixed approach, intended to combine human and artificial intelligences, which means that the process assisted teachers identifying students' profile, allowing the creation of personalized recommendations. We need more experiments with different scenarios, but this evidence suggests the PDMP guided pedagogical decision-making, meeting our first research objective.

It is important to notice that both the amounts of problems solved correctly and of the problems solved incorrectly, decreased after the personalized recommendations. This can be due to the fact that most learners, normally, accessed the learning environment to solved problems, only. Teachers recommended them to interact with the aspects of the environment they were neglecting (social and collaborative aspects). Because of that, they spent less time solving problems. Another important result to notice is that the difference on the average between the problems solved correctly and those solved incorrectly, decreased as well (from 6.77 to 4.33), which is desired and means learners made less mistakes while solving problems.

5.2 RO2: Create an Authoring Solution, Based on the Process, to Build Informed Pedagogical Decisions, in Course Time, for Online Learning Environments

This was part of experiment 2, where we asked participants to accomplish some tasks that, in the end, should create a *Pedagogical Decision Capsule*. We measured the participants precision in correctly accomplishing the tasks and normalized the results. In order to do that, participants would need to understand the situation they were dealing with, which is displayed by the visualizations. The formula is presented below:

[10] The learners avoiding problem solving were recommended to do so, that is why we saw an increase in problem solving (both correctly and incorrectly), which means learners were doing what they were asked to.

$$precision = \frac{\sum_{i=1}^{n} e_i}{MAX_{step}}$$

In Table 2, we compare the authoring versions (HW and LW), regarding the precision on accomplishing each task (for each step).

Table 2. Wilcoxon-Mann-Whitney test - precision

Steps	P-value	Greater precision
1	0.0001984*	*HW*
2	0.08513	No difference
3	0.01698*	*HW*
4	0.0004147*	*LW*
All steps	0,02922*	*HW*

The conclusion is that the HW version promoted greater precision in the accomplishment of the tasks. The reasons are not clear, but we believe that the greater control that instructors had using the HW version, helped them accomplish the proposed tasks. In future works, we will study the reasons for such outcome.

5.3 RO3: Enable the Authoring of Pedagogical Decisions at Different Levels of Granularity Without Affecting the Users' Perception About the Authoring Tool

The objective is to evaluate if the authoring solution allows pedagogical decisions to be made at different levels of granularity (different versions of the authoring solution). We evaluated the interface considering the users': (1) utility perception - UP; (2) ease of use perception - EUP; (3) attitude towards use - ATU; (4) intention to use - IU. Participants (users) should assign a score (Likert scale from 0 to 6) for each criteria.

In Table 3, we summarized the average of the scores for each version (heavy weight - HW, light weight - LW, spreadsheet - SS):

Table 3. Average scores of the participants' perceptions.

	UP	EUP	ATU	IU
HW	3.9	3.8	3.8	4.1
LW	4.1	3.9	4.0	4.2
SS	3.7	4.1	3.7	3.9

Regarding **utility perception**, **ease of use perception**, **intention to use**, and **attitude towards use**, we can conclude that there is a statistical similarity between the three versions used in the experiment[11].

Analyzing the responses, we find that the mean of the responses are higher or very close to 4 ("I somewhat agree") for all metrics, at all stages and for all versions. This result indicates that the participants' perceptions were positive and similar, concluding that the authoring solution allows pedagogical decisions to be made at different levels of granularity without affecting the users' perception of authorship.

5.4 RO4: Define Visual Metaphors to Allow Instructors, Transparently, to Use Techniques of Educational Data Processing and Analysis, to Inform Them of Relevant Pedagogical Situations Occurring During Course Time

The objective is to evaluate if the visualizations used helped instructors understand the result of the data analysis. In order to meet this objective, we analyzed if participants correctly answered the questions in step 2 (where the visualizations were presented to assist participants). We also analyzed the participants' perceptions regarding the *use of RAG colors*[12] and the *terms used to refer to the learners' performance*[13].

We created 4 visualizations. One was used in the *heavy weight* version (Viz1) and the other 3 were used in the *light weight* version (Viz2, Viz3 and Viz4). In Table 4, we compared the visualizations among themselves, considering how well they helped instructors understand the results from the educational data analysis. We applied the Bonferroni correction because several independent statistical tests are being performed, simultaneously, on the same data set.

The results show that Viz1 and Viz2 promoted greater correctness in the participants' answers to the tasks (i.e.: helped instructors understand the result of the data analysis). The results show the following order in terms of assistance brought by the visualizations: Viz1 > Viz2 > Viz3 = Viz4.

Considering the *use of RAG colors*, we compared the version (Table 5).

The results for the perception on the use of RAG Colors was not significantly (statistically) different. This is the desired result, the use of RAG Colors is a resource to support instructors and should, therefore, be consistently provided in all authoring versions.

Finally, considering the *terms used to refer to the learners' performance*, the results are shown in Table 6.

[11] There was a small difference in step 4, regarding the utility perception and the intention to use.

[12] We used the colors Red, Amber and Green to categorize learners according to their performance/pedagogical situation (inadequate: red; insufficient: amber/yellow and adequate: green).

[13] Inadequate, insufficient and adequate.

Table 4. Comparison between visualizations.

VIZ	Wilcoxon-Mann-Whitney	Bonferroni	Greater correctness
Viz2 Viz3	0.0001185024	0.0007110142*	Viz2
Viz2 Viz4	7.216746e-06	4.330048e-05*	Viz2
Viz2 Viz1	1.00		No difference
Viz3 Viz4	0.01171951	0.07031707	No difference
Viz3 Viz1	3.330479e-05	0.0001998287*	Viz1
Viz4 Viz1	1.439644e-06	5.758577e-06*	Viz1

Table 5. Wilcoxon-Mann-Whitney test, comparing the participants' perception on the use of RAG Colors

Wilcoxon-Mann-Whitney		
Steps	P-value	Greater perception
1	0.5728	No difference
2	0.07699	No difference
3	0.1998	No difference
4	0.07501	No difference

Table 6. Wilcoxon-Mann-Whitney test, comparing the participants' perception on terms used to refer to the learners' performance

Wilcoxon-Mann-Whitney		
Step	P-value	Greater perception
1	0.4632	No difference
2	0.1593	No difference
3	0.09437	No difference
4	0.0196	*LW*
All	0.02744	*LW*

In this case, the way the terms were used in the *light weight* version were (significantly) better perceived by the participants. This result is not aligned with our expectations, demanding future improvements regarding this matter.

6 Conclusion

In this chapter, we presented data showing the growing interest in distance education, especially with regard to MOOCs (Massive and Open Online Courses). We presented the problem that instructors face when trying to deal with the large amount and diversity of MOOC data, highlighting the impossibility of manually obtaining relevant information in course time.

We have proposed a process, called the Pedagogical Decision-Making Process (PDMP), as well as an authoring solution to help teachers and tutors follow the PDMP, without the need for advanced technical knowledge, to make pedagogical decisions informed by educational data (T-Partner).

To evaluate our proposals, we performed 2 experiments, using educational data from a MOOC (MeuTutor-ENEM). The first was to evaluate the process, where we compared the effectiveness of personalized pedagogical recommendations following the PDMP, with pedagogical decisions automatically generated by the mentioned MOOC. The results showed that the recommendations following the PDMP improved students' interactions.

The second experiment was to evaluate if the authoring solution could help instructors, supporting pedagogical decision-making (at different levels of granularity and with the use of special colors scheme, RAG colors; data visualization and appropriate terms/expressions). The results demonstrate that the participants were able to perform the proposed tasks. Participants' responses indicated a positive perception regarding all the metrics and we found no statistically significant difference between the 3 versions.

Although we noticed some parts that need adjustment, the results, in general, showed a positive perception of the participants, with small differences among the experimented versions. We can conclude that results were positive and met our expectations.

References

1. Baker, R.S.: Stupid tutoring systems, intelligent humans. Int. J. Artif. Intell. Educ. **26**(2), 600–614 (2016)
2. Belanger, Y., Thornton, J.: Bioelectricity: A Quantitative Approach, Duke University's First MOOC (2013)
3. Bienkowski, M., Feng, M., Means, B.: Enhancing teaching and learning through educational data mining and learning analytics: an issue brief, US Department of Education, Office of Educational Technology, pp. 1–57 (2012)
4. Bittencourt, I.I., Costa, E., Silva, M., Soares, E.: A computational model for developing semantic web-based educational systems. Knowl.-Based Syst. **22**(4), 302–315 (2009)
5. Cabada, R.Z., Estrada, M.L.B., García, C.A.R.: Educa: a web 2.0 authoring tool for developing adaptive and intelligent tutoring systems using a Kohonen network. Expert Syst. Appl. **38**(8), 9522–9529 (2011)
6. Chou, C.-Y., Huang, B.-H., Lin, C.-J.: Complementary machine intelligence and human intelligence in virtual teaching assistant for tutoring program tracing. Comput. Educ. **57**(4), 2303–2312 (2011)

7. Chrysafiadi, K., Virvou, M.: Student modeling approaches: a literature review for the last decade. Expert Syst. Appl. **40**(11), 4715–4729 (2013)
8. De Bra, P., et al.: ALAT: finally an easy to use adaptation authoring tool. In: Proceedings of the 27th ACM Conference on Hypertext and Social Media, pp. 213–218. ACM (2016)
9. Associação Brasileira de Educação á Distância. Censo ead br (2015). Relatório Analítico da Aprendizagem a Distância no Brasil, 2016
10. Dermeval, D., Paiva, R., Bittencourt, I.I., Vassileva, J., Borges, D.: Authoring tools for designing intelligent tutoring systems: a systematic review of the literature. Int. J. Artif. Intell. Educ. 1–49
11. Foss, J.G.K., Cristea, A.I.: The next generation authoring adaptive hypermedia: using and evaluating the MOT3. 0 and PEAL tools. In: Proceedings of the 21st ACM Conference on Hypertext and Hypermedia, pp. 83–92. ACM (2010)
12. The William Flora Hewlett Foundation: White paper: Open educational resources. breaking the lockbox on education. Technical report, The William Flora Hewlett Foundation (2013)
13. Heffernan, N.T., Heffernan, C.L.: The assistments ecosystem: building a platform that brings scientists and teachers together for minimally invasive research on human learning and teaching. Int. J. Artif. Intell. Educ. **24**(4), 470–497 (2014)
14. Jain, A.K.: Data Clustering: 50 Years Beyond K-means. Pattern Recogn. Lett. **31**(8), 651–666 (2010)
15. Kopcha, T.J.: Teachers' perceptions of the barriers to technology integration and practices with technology under situated professional development. Comput. Educ. **59**(4), 1109–1121 (2012)
16. Liyanagunawardena, T.R., Parslow, P., Williams, S.: Dropout: MOOc participants' perspective (2014)
17. Murray, T.: Authoring intelligent tutoring systems: an analysis of the state of the art. Int. J. Artif. Intell. Educ. (IJAIED) **10**, 98–129 (1999)
18. Murray, T.: An overview of intelligent tutoring system authoring tools: updated analysis of the state of the art. In: Murray, T., Blessing, S.B., Ainsworth, S. (eds.) Authoring Tools for Advanced Technology Learning Environments, pp. 491–544. Springer, Heidelberg (2003). https://doi.org/10.1007/978-94-017-0819-7_17
19. Onah, D.F.O., Sinclair, J., Boyatt, R.: Dropout rates of massive open online courses: behavioural patterns. In: EDULEARN14 Proceedings, pp. 5825–5834 (2014)
20. Paiva, R., Barbosa, A., Batista, E., Pimentel, D., Bittencourt, I.I.: Badges and XP: an observational study about learning. In: Frontiers in Education Conference (FIE), 2015. 32614 2015. IEEE, pp. 1–8. IEEE (2015)
21. Paiva, R., Bittencourt, I.I., da Silva, A.P.: Uma ferramenta para recomendação pedagógica baseada em mineração de dados educacionais. In: Anais dos Workshops do Congresso Brasileiro de Informática na Educação, vol. 2 (2013)
22. Paiva, R., Bittencourt, I.I., Tenório, T., Jaques, P., Isotani, S.: What do students do on-line? Modeling students' interactions to improve their learning experience. Comput. Hum. Behav. **64**, 769–781 (2016)
23. Paiva, R.O.A., Bittencourt, I.I., da Silva, A.P., Isotani, S., Jaques, P.: Improving pedagogical recommendations by classifying students according to their interactional behavior in a gamified learning environment. In: Proceedings of the 30th Annual ACM Symposium on Applied Computing, pp. 233–238. ACM (2015)
24. Romero, C., Ventura, S.: Educational data mining: a review of the state of the art. IEEE Trans. Syst. Man Cybern. Part C: Appl. Rev. **40**(6), 601–618 (2010)

25. Romero, C., Ventura, S.: Educational data science in massive open online courses. Wiley Interdisciplinary Rev. Data Mining Knowl. Discov. (2016)
26. Schildkamp, K., Lai, M.K., Earl, L. (eds.): Data-Based Decision Making in Education: Challenges and Opportunities, vol. 17. Springer, Heidelberg (2012). https://doi.org/10.1007/978-94-007-4816-3
27. Siemens, G., d Baker, R.S.J.: Learning analytics and educational data mining: towards communication and collaboration. In: Proceedings of the 2nd International Conference on Learning Analytics and Knowledge, pp. 252–254. ACM (2012)
28. Bittencourt, I.I., et al.: Evaluating the impact of Mars and Venus effect on the use of an adaptive learning technology for Portuguese and Mathematics. In: The 16th IEEE International Conference on Advanced Learning Technologies, ICALT 2016 (2016)
29. Sottilare, R., Graesser, A., Hu, X., Brawner, K.: Design Recommendations for Intelligent Tutoring Systems: Authoring Tools and Expert Modeling Techniques (2015). Ed by Robert Sottilare
30. Teo, T.: A path analysis of pre-service teachers' attitudes to computer use: applying and extending the technology acceptance model in an educational context. Interact. Learn. Environ. **18**(1), 65–79 (2010)
31. Teo, T.: Factors influencing teachers' intention to use technology: model development and test. Comput. Educ. **57**(4), 2432–2440 (2011)
32. Teo, T., Luan, W.S., Sing, C.C.: A cross-cultural examination of the intention to use technology between Singaporean and Malaysian pre-service teachers: an application of the technology acceptance model (TAM). Educ. Technol. Soc. **11**(4), 265–280 (2008)

User Modelling and Grouping

Group Formation in CSCL: A Review of the State of the Art

Simone Borges[1(✉)], Riichiro Mizoguchi[2], Ig Ibert Bittencourt[3], and Seiji Isotani[1]

[1] University of São Paulo, São Carlos, SP, Brazil
{sborges, sisotani}@icmc.usp.br
[2] Japan Advanced Institute of Science and Technology, Nomi, Kanazawa, Japan
mizo@jaist.ac.jp
[3] Federal University of Alagoas, Maceió, AL, Brazil
ig.ibert@ic.ufal.br

Abstract. Group formation in CSCL refers to the process of adequate selection and grouping of students to create scenarios and situations that help the learning to occur more smoothly. Objective: this paper presents a systematic mapping of the literature about group formation for CSCL intended to characterize the state of the art in the field as well as identifying gaps and opportunities for further re-search. Method: We designed a protocol to collect and analyze the literature on group formation in CSCL and carried it out using a rigorous systematic re-view/mapping method established in the literature. Initially, we collected 3571 papers about CSCL that had the potential to provide important information about research on group formation. After initial screening, 423 were recognized as papers related to group formation. After a careful analysis, 106 papers met the necessary requirements/criteria defined in our protocol. Results: each of the 106 papers was categorized according to their contributions using information extracted from this systematic mapping, and a framework to classify research in the field is proposed. Conclusions: this work provides an extensive analysis of the literature on group formation for CSCL offering an overview of the state of the art as well as opportunities for future research. We also create an infographic to summarize our findings, available at http://infografico.caed-lab.com/mapping/gfc/.

Keywords: Systematic mapping · Group formation · Literature review
Computer-Supported Collaborative Learning (CSCL)

1 Introduction

Collaborative learning (CL) has been discussed for the last two decades, although recently gained prominence because of advances in computer technology (e.g. widespread use of tablets and mobile phones) and the widespread use of Internet-based educational tools (e.g. LMSs and MOOCs). As pointed out by [29] the Internet has brought up renewed interest in investigating the benefits of computer support when applied to CL. These advances have set the stage for Computer-Supported Collaborative Learning (CSCL), which is a pedagogical strategy where individuals study in

A. I. Cristea et al. (Eds.): HEFA 2017, CCIS 832, pp. 71–88, 2018.
https://doi.org/10.1007/978-3-319-97934-2_5

groups, and knowledge is gained and constructed through discussions, argumentations, exchange of ideas, conflict resolution, and so on. In such a pedagogical strategy, computers play a key role in supporting the learning process by making feasible the design, orchestration and assessment of group activities in several different contexts and situations [28, 29]. According to the findings of the research community, collaboration has a positive influence on the learning process, mainly when it is well designed and properly introduced. Therefore, a CSCL environment and CL activities should be carefully designed; otherwise there is no guarantee that the learning results will meet students' and teachers' expectations. In this context, group formation plays a fundamental role. It influences how students perceive the environment, interact with peers, use available didactic materials, and take part in learning activities and processes [20].

Group formation in CSCL refers to strategies, algorithms, techniques and methods to cluster individuals according to several criteria with the objective of creating well-thought-out groups that will lead students to better interact with each other and maximize their learning gains. Thus, when a group is adequately created, every student brings something relevant to the table, adding improvements to the group learning experience. With the intention to proper selection students to create effective groups, characteristics such as: knowledge level, learning style, grades, and language barriers, to name a few, have been considered in hopes of improving learning interactions and benefits [19]. Although many studies have investigated the benefits of group formation in CSCL to date, few studies have analyzed the achievements of the research published on the topic and categorized them according to their contribution. For example, to the best of our knowledge, there is no study summarizing the main group formation approaches, the key student's characteristics taken into account during group formation, educational level in which the approaches are meant to be applied, most common learning activities, and so on. To fill such a gap and provide an overview of this research area, we carried out a process to systematically review the literature, known as systematic mapping process [12, 13]. It requires creating a protocol to systematically search and analyze published articles of a given research area.

Upon carrying out this systematic mapping, we went through 3571 papers about CSCL, among them, 423 were recognized as papers related to group formation, and only 106 met the criteria for inclusion of articles defined in this study. These 106 papers were selected and categorized according to their contribution. It is important to say that in this paper, references in the format (authors, year) refer to papers included in the systematic mapping, due space restrictions, a list of these papers is available at https://goo.gl/V8cbEo. In particular, this work intends to answer the following questions regarding the research on group formation in CSCL: (i) What are the most investigated characteristics of group formation in the domain of CSCL? (ii) At what educational level group formation has been most investigated and applied to? (iii) What research methods have been most used for evaluating the formation of groups? (iv) What are the main venues in which research in this area has been published? The paper is organized as follows. Section 2 provides a background and a framework to classify and understand the research on group formation. Section 3 details the selected primary studies and presents an analysis of these studies. It also further analyzes the results by describing the most widely investigated topics on group formation and elaborates on their practical implications. Section 5 covers threats to validity, and Sect. 6 presents concluding remarks.

2 Framework to Classify and Understand Research on Group Formation

There are several denominations for the act of grouping learners in the context of collaborative learning, such as, group/team composition [17] group/team formation [33], grouping students [3], clustering students [8]. These denominations have the same meaning, as the act of bringing together students in effective learning groups [19]. In this work, we adopt the term group formation, which is the most widely term used in the investigated studies. Although the term "effective" can have different meanings among researchers, in this field is often used as a synonym for the proper allocation of resources to enhance the learning process [19]. These resources can be tangible (e.g. learning materials and tools to support collaboration) or intangible (e.g. knowledge and skills to be learned). Most research on group formation have focused on how to allocate these resources based on the learners' profiles, on the technologies involved, or on the tasks to be performed (CL techniques, also known as CL best practices) [23].

Using learners' profiles helps instructors adequately deliver custom content to satisfy, not only individual necessities of the learners, but also group's necessities [2, 3, 14, 15, 27], while inputs from the environment, can give more information that can help to understand the collaborative context and as a result, to improve grouping quality [16, 25, 30, 31]. Finally, there are other approaches for group formation that relies on using specifications of CL techniques to acquire sufficient information to group the learners [8]. Generally, the formation of groups can be performed randomly (e.g. assigning learners in the groups by chance), self-selected (i.e. the learners choose with whom they want to study), or it can be carried out by an instructor or computing system, based on some criteria [2, 3, 7, 14–18, 27].

As discussed by [10, 22], the benefits of collaborative learning arise only when there is a detailed and proper planning of components and collaborative mechanisms. Considering these, many authors have raised concerns regard random selection and self-selected approaches since these approaches can result in unequal participation such as, students in the same group working at a different pace, off-task behavior, or even increasing students' resistance to group work [6, 9, 20]. Moreover, one way to increase the chances of forming groups more capable to achieve the desired learning goals is assigning learners to groups using approaches based on CL techniques [6, 8, 9, 29]. As discussed in the previous paragraphs, several topics have been investigated under the umbrella of group formation research. Thus, to understand the contribution of the community we propose a framework based on our understanding of the field. This framework (Fig. 1) is composed of six main categories and their relationships:

1. **Planning:** The formation of groups can either follow a set of criteria (i.e., systematic group formation) or happen randomly (i.e., unsystematic group formation). The criteria define the conditions that govern group formation, and the clustering algorithms determine how students should be selected and grouped together.
2. **Initiative:** to form groups we define two types of initiatives: spontaneous and controlled. Spontaneous group formation uses only the preferences of students to forming groups. This allows for the formation of like-minded groups of students. Controlled group formation, on the other hand, entails putting forward criteria and algorithms aimed at guiding the formation of groups. It might lead to heterogeneous groups in hopes of taking advantage of the resulting diversity of backgrounds as well as enhancing the overall learning experience.
3. **Diversity of population:** Given each particular criterion, the population diversity in a group can be considered either homogeneous (i.e., whose participants have the same or similar trait) or heterogeneous to various degrees (i.e., participants differ from each other in at least one trait), depending on the number of distinct individuals in the group regard the considered criterion.
4. **Distribution approach:** Despite how many criteria a group formation strategy uses, if students end up distributed either in homogeneous or heterogeneous groups, we named the distribution approach as being *Simple*. However, when more than one criterion is used, the distribution of students can result in one (or more) criterion used to group the students homogeneously, while another different criterion (or criteria) being used to distribute the students heterogeneously. As a result, these groups can be considered homogeneous and heterogeneous at the same time, given the considered criterion. We named such distribution approach as *Hybrid* distribution.
5. **Computer-supported:** group formation can be very simple by using only one criterion (e.g. gender) or quite complex (using several criteria and algorithms). Thus, it is possible to categorize studies whose group formation approach relies on computational support and studies that report on approaches that do not rely on any sort of automated support for group formation.
6. **Rationale:** studies can also be classified in terms of how they back up their approach for group formation. In this framework we classify three types: studies that do not present any arguments that back up their approaches, studies whose group formation approach is borne out by empirical evidence, and studies that draw on assumptions from pedagogical theories.

The categories are connected in the framework according to the following rationale: in general, the formation of groups is based on information (or lack of information) on the learner profiles. (1) The formation of groups can be planned or being randomly carried out. In the case of being planned, it can be based on grouping any criteria. Several criteria can be used, such as information about the students, students' preferences, or even constraints of the system or environment. (2) When students decide the clustering process, the initiative is spontaneous. When students do not participate in the selection process, and any other criterion is used to distribute them in groups, group formation is called controlled. (3) Criteria can be used both to group students with bases in their similarities (homogeneous groups) and/or in their differences (heterogeneous groups). (4) In simple distribution, the groups are mutually exclusive, being either homogeneous or heterogeneous; however, there are hybrid approaches in which groups are, at the same time, homogeneous and heterogeneous. (5) The formation of groups can be done manually, or it can rely, wholly or partially on computational systems. (6) Finally, the rationale provided in the studies allows us to understand whether the study is based on any pedagogical approach or based on empirical evidence; yet it is still possible to find studies which neither one nor another kind of justification for the formation of groups is provided.

Figure 1 gives an overview of our framework that can be used to classify, analyze and understand different types of research on the topic. The categories in the framework are tailored to give an overview of the efforts that have been carried out in the area, support the comprehension of the most investigated topics, and highlight the interacting and overlapping topics. Nevertheless, it is worth emphasizing the categories we came up with are not exhaustive. That is, they do not cover all the dimensions of the research spectrum of group formation in the context of CSCL. In addition, we created three categories not showed in Fig. 1. The purpose of these categories is to support gathering information about the way the selected researches have being conducted, and not directly about the group formation process.

- **Educational level** (or learning activities)**:** By analyzing, in which in educational contexts group formation has been explored, one can determine research gaps and future research opportunities;
- **Research methods:** This category characterizes the type of research that has been carried out and reported in the studies;
- **Publication type:** As for the publication types, we have selected primary studies that have been published in conferences, journals, and workshops.

2.1 Method: Systematic Mapping

Thus, obtaining an overview of an established research area can be a complex undertaking. Evidence-based Software Engineering [13] proposes a set of guidelines to support the conduction this sort of investigation. More specifically, these guidelines outline how to identify, evaluate, interpret, and analyze primary studies [12]. The main assumption is that those guidelines, which are expressed and embodied in systematic mappings and systematic reviews, lead to more consistent results that can be more easily replicated. Due to space restriction, the complete systematic mapping protocol can be found at: https://goo.gl/V8cbEo.

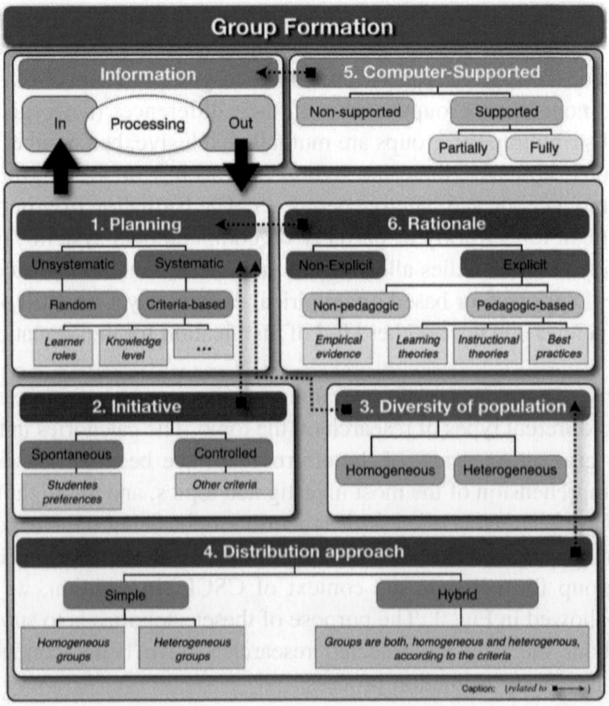

Fig. 1. Overview of the framework to classify and understand research on group formation.

3 Systematic Mapping Results

In this section, we analyze the results of our systematic mapping. We draw information from the primary studies in order to answer the RQs of our mapping study.

3.1 Research Question 1: What are the Most Investigated Characteristics of Group Formation in the Domain of CSCL?

The categories, as mentioned earlier, were created with the purpose of shedding some light on what are the most investigated characteristics of group formation in the domain of CSCL. By characteristics we mean that we applied a *keywording* strategy to prepare our rating system and characteristics for the selected primary studies. By applying such a strategy, initially summaries were read to find keywords and concepts that reflect the study's contribution. Next, these keywords and concepts were combined to produce a general understanding of the nature and contributions of the research. The classification scheme gradually evolved towards its final version as some characteristics were eliminated, added, merged or split [11]. In addition, grouping the primary studies into categories allowed us better to grasp the interrelationships and overlaps on research efforts in the field. However, it is worth mentioning that the resulting categories do not encompass all characteristics, research themes and contributions in the area. Rather, the categories were devised in hopes of better answering the RQ1.

Group Formation Planning. As mentioned, this category groups the primary studies in terms of the sort of planning that was used during group formation. Here we classify as unsystematic group formation approaches within which the formation of groups lacks a set of governing rules and takes places in a seemingly random fashion. In contrast to unsystematic group formation, systematic group formation is based on criteria that govern the size and organization of groups. In Fig. 1, the box labeled 1-Planning shows the proposed classification for this category, highlighting the basic difference between systematic and unsystematic group formation. In 12 of the selected studies, students are grouped in a random fashion. Nevertheless, in only 6 of these studies randomness is the only factor determining group formation. In the other studies, it was also possible to group students based on criteria aimed at achieving the target educational objectives. As an example, in the study conducted by [26], initially the groups were unsystematic formed (i.e. by chance), given that not enough information to warrant the application of more complex criteria about the students was collected yet. Later, after the conduction of more activities, the students were evaluated, and the results of the evaluations were then used as a group formation criterion. According to [26], this approach could be an alternative to using placement or pre-evaluation questionnaires. One drawback of using questionnaires before collaborative learning activities is that carrying out this sort of evaluation is a time consuming task, and it is often seen as tedious for the students taking part in it. Most primary studies report on group formation approaches that are based on some criteria: in 94 studies, criteria are used to support the definition of rules that establish which group each student should be added to. Figure 2 shows the frequency of the criteria for the formation of groups found in primary studies analyzed. The quantity of criteria present in each study varies according to the objectives of the collaborative activities to be performed or can be based on the kind of computational support employed (e.g. algorithms). Abnar et al. [1] remark that, in general, when groups are manually created, only one criterion is used.

The most used criterion for group formation was knowledge level. Knowledge level-based group formation is investigated in 59 primary studies. Skill-based group formation is the second most investigated criterion, investigated in 33 primary studies. Learner roles are the third most investigated criteria, 31 primary studies. As shown in Fig. 2, the other criteria that have been explored are group objectives, individual objectives, student's genre, cultural aspects (e.g., nationality and language), interaction issues, and characteristics extracted from social relationships, psychological characteristics, learning disabilities, schedule compatibility, age, and academic formation. It is worth mentioning that these are only some of the identified criteria, studies might have used other additional criteria. However, these other criteria were not mapped because the studies failed to describe sufficiently them. Usually, these additional criteria are simply mentioned in the primary studies as "other criteria" or "among other things".

Initiative in Systematic Group Formation. When group formation follows a methodical (viz., systematic) approach to group formation, the initiative for group formation can be handled in two different ways: spontaneous group formation (i.e., when the students are the ones in charge, yielding self-selected groups) and controlled group formation (i.e., when the instructor or underlying approach controls group formation, resulting in instructor-select groups). This distinction is highlighted in Fig. 1, box 2-Initiative. We classify as spontaneous the approaches in which students are free to form their own groups. Spontaneous group formation approaches take into account the preferences of the students (e.g., affinity between students, idiosyncrasies, and friendship). Therefore, we classified as controlled group formation (guided) the approaches that take some other elements into account during the formation of groups, apart from students' preferences. We argue that the term controlled better describe such approaches, given that there is an external element, which can be impervious to the students' preferences, effectuating the rules and controlling how students should be grouped.

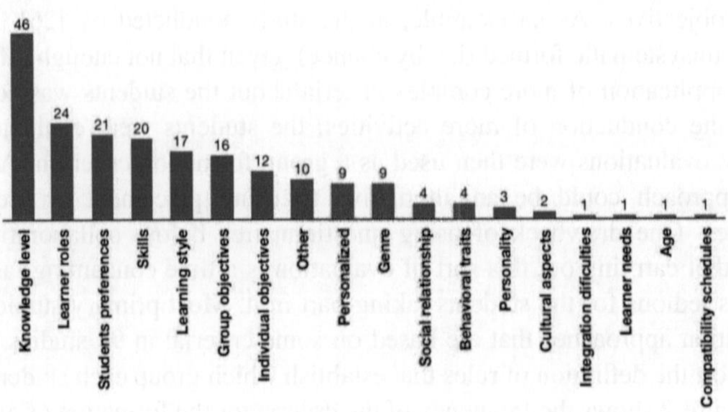

Fig. 2. Frequency of the criteria for the formation of groups found in primary studies analyzed.

In 21 of the selected studies, the preference of the studies was taken into account during group formation. Nevertheless, in only four studies this was the only factor taken into consideration. In the other 17 studies, apart from the students' preferences, other criteria could also be used. In these primary studies, the students are polled about their preferences, and the gathered information is then used to devise group formation criteria. The Venn diagram shown in Fig. 3 illustrates how the primary studies were categorized in terms of their planning strategy. As mentioned, some primary studies can fall in more than one category since their approaches are flexible enough to allow for more than one planning strategy. By analyzing Fig. 3, it can be seen that most group formation approaches (70 primary studies) are exclusively criteria-based approaches.

Population Diversity. In the context of this study, diversity has to do with the makeup of groups, which can be *heterogeneous* or *homogeneous* as shown in Fig. 1, box 3-Population diversity. Homogeneous approaches emphasize criteria that result in groups of like-minded students, i.e., students that present similar characteristics. Heterogeneous approaches, conversely, propose the creation of groups that are comprised of members with different backgrounds and characteristics. The most used approach is the formation of heterogeneous groups (84 primary studies). Evidence suggests that heterogeneous groups foster the interaction among group members due to the members have skills that are complementary to the group as a whole [21, 24]. Also, in heterogeneous groups it is normal to have members with characteristics that supplement other individuals in the group.

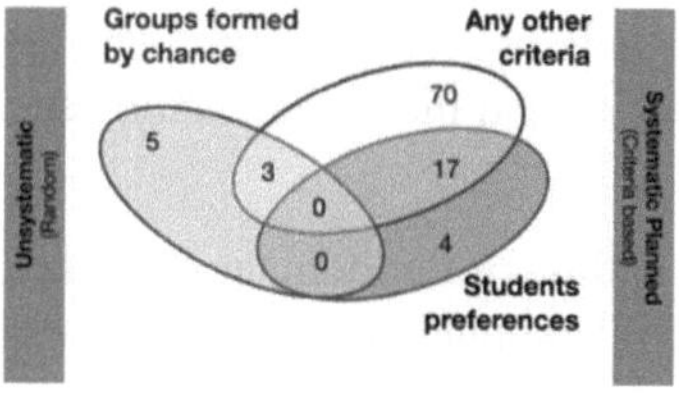

Fig. 3. Distribution of the primary studies according to the clustering strategy.

In 36 primary studies relied on the formation of homogeneous groups. Note that in two of these 36 studies the formation of homogeneous groups is the only option, the other 34 studies also propose criteria that yield heterogeneous groups (Fig. 4). In addition, nine papers were not classified in terms of diversity, given that in five of them the formation of groups can be carried out in a random fashion and four in a spontaneous way. Apart from these papers, we also analyzed 17 primary studies that either use or investigate group formation. However, it was not possible to identify the diversity of the groups in these papers.

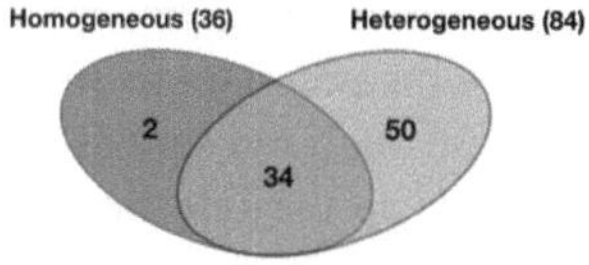

Fig. 4. Primary studies according to diversity of the population in the groups.

In the study conducted by Chiu and Hsiao [8], it is argued that randomly assembled groups can result in an acceptable level of heterogeneity with respect to certain characteristics as, for instance, the number of students of the same gender or with some particular personality traits (e.g., active students, passive students, leaders, and followers). Moreover, in the study conducted by Ardaiz-Villanueva et al. [4], problem-based learning (PBL) principles and several support tools were used to support the conduction of collaborative learning activities. Two criteria drive group formation: the first is based on a "creativity score" attributed to each student and the second boils down to computing an "affinity score" that gauges how each student gets along with others. According to the authors, the results seem to suggest that homogeneous group (in terms of creativity) lead to a better learning experience.

Distribution Approaches. We categorized the studies according to the way the students were distributed in groups. As mentioned, groups can be homogeneous or heterogeneous groups according to the criterion (or criteria) used. In terms of distribution, we classified the studies whose approaches result in either homogeneous or heterogeneous groups (despite how many criteria were used) as *simple distribution*. By the other hand, we also found studies in which a subset of criteria was used to cluster the students in homogeneous groups, while at the same time; another subset of criteria was used to cluster them in heterogeneous groups. This second type of approach was classified as *hybrid distribution*, and the classification scheme is shown in Fig. 1, box 4-Distribution approach.

Computational Support. Although the domain of computer-supported collaborative learning contains the words "computer-supported", not always the group formation occurs with technological support. Based on these observations, we devise the category shown in Fig. 1, box 5 - Computer-Supported, to classify the primary studies according the computational support. We observed that a few primary studies (13) did not provide enough information clarifying neither the use nor the extension of the computational support, getting this restricted to the other stages of collaborative learning (i.e. collaborative tasks to be performed). In the remaining 83 studies analyzed, various technological solutions have been used. Although the extent and the way these technologies were employed are relevant, they are beyond the scope of this paper and will not be detailed. Due space restrictions, more information is available at https://goo.gl/V8cbEo.

Rationale Behind the Strategy for Group Formation. The primary studies were also classified in terms of how the authors back up their approach for group formation. We found three types of studies: (i) studies that do not present any arguments that back up their approaches, (ii) studies whose group formation approach is borne out by empirical evidence, (iii) and studies that draw on assumptions from pedagogical theories (Fig. 1, box 6-Rationale). Only 37 of the 106 analyzed studies are based on sound pedagogical or instructional theories, which back up the selection criteria employed during group formation. We also found that 46 studies do not report whether any pedagogical theory was used to support the group formation criteria. These 46 studies mention, however, that the chosen criteria are based on empirical evidence. Apart from these primary studies, 23 studies do not elaborate on what theories ground their criteria.

The Answer for RQ1 (What are the Most Investigated Characteristics of Group Formation in the Domain of CSCL?). The majority of the primary studies (101 out of 106) report on approaches for planned group formation, i.e., the formation of groups is based on pre-established criteria. The most widely used criterion is the level of knowledge of the students, which is used in 46 of the selected studies. Controlled group formation is used in 90 primary studies. In terms of distribution, most primary studies rely on heterogeneous groups: 84 studies propose the formation of heterogeneous groups. Nevertheless, only in 50 out of these 84 studies simple heterogeneous groups are the only possible outcome. According to our results, 82 studies capitalize on some sort of technology to automate and support group formation. Only 37 studies exploit educational or pedagogical theories to justify their rationale for the group formation approach.

3.2 Research Question 2: In What Educational Level (or Learning Activities) Group Formation has been Most Investigated and Applied to?

In order to answer RQ2, we classified the selected studies in terms of their target population. In the context of this paper, target population refers to the stage (or level) at which educational activities take place or the type of educational activities. As shown in Fig. 5, the group formation approaches proposed in most studies (56) emphasize tertiary education (viz., undergrad students). Few studies focus on applying group formation approaches to other educational levels (or stages). For instance, our results would seem to suggest that there has been scant research on the impact of group formation when applied to early childhood education (1), graduate students (1), e-Health students (1), language-learning activities (1), primary education students (3), and secondary education students (6). 36 primary studies fail to specify the target population of their proposed approach. In addition, these studies present theoretical models that represent solutions that have not been implemented or fully tested yet.

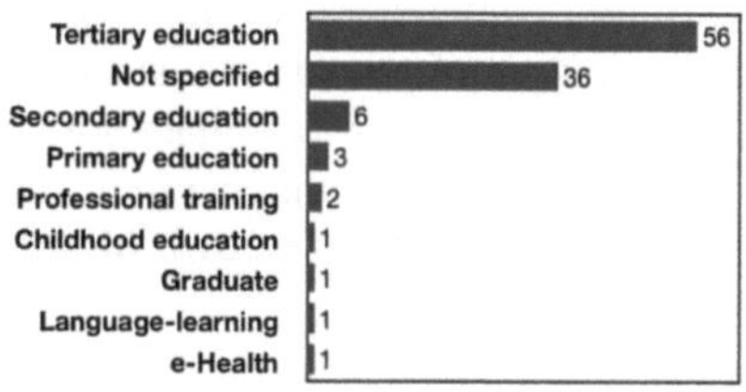

Fig. 5. Distribution of primary studies according to the level/type of educational activity.

The Answer for RQ2 (In What Educational Level (or Learning Activities) Group Formation has been Most Investigated and Applied To?). As mentioned in the previous section, most research on group formation applied to CSCL has been geared towards undergrad students: approximately 52% (56 primary studies) of the studies tackle group formation at this educational stage (level).

3.3 Research Question 3: What Type of Research is the Most Common in the Field?

The classification proposed in [32] comprises the following research types: Validation research, Evaluating, Philosophical papers, Position papers, Solution proposals, Experience papers, and Tools. Solution proposals are by far the most common research method, including 77 studies, 17 primary studies are concerned with some evaluation, and only one study falls into the opinion category, as shown in Table 1. Moreover, the main contribution presented by 11 studies boils down to the implementation of a tool to automate and support group formation, falling into the tools category.

Table 1. Distribution of primary studies by research type.

Research type	Solution proposal	Evaluation	Tool	Opinion
Quantity	77	17	11	1

The Answer for RQ3 (What Type of Research is the Most Common in the Field?). As discussed in the previous section, the research method used in the majority of the primary papers is solution proposal: 77 studies present group formation approaches that can be applied to CSCL environments. Conversely, the least explored category is opinion. Only one study describes the author's view about group formation applied to CSCL environments and the implications thereof.

3.4 Research Question 4: What are the Main Venues in Which Research in this Area has been Published?

As shown in Fig. 6-left, most primary studies are indexed in IEEE (IEEE Xplore) and Elsevier (26 and 24 papers, respectively). The selected papers were published in a variety of venues that range from conferences, workshops, to journals. Figure 6-right shows an overview of the distribution of primary studies according to their types over the years. By analyzing Fig. 6-left, we can see that most efforts in the area were published from 2006 on. Since then, it seems that the topic has gained momentum. As indicated in Fig. 6-right, there was a marked increase in the number of journal publications in 2011.

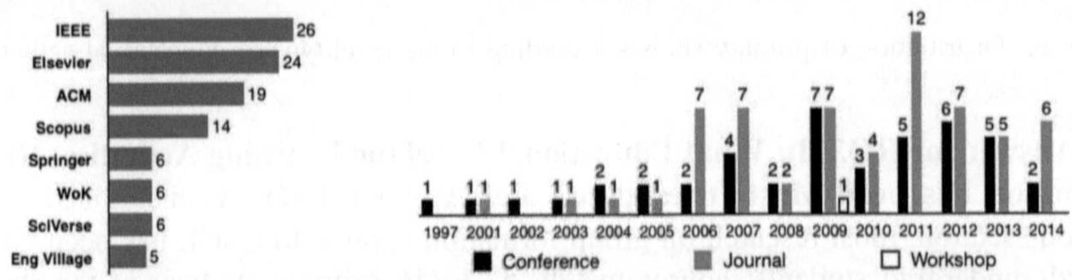

Fig. 6. Left: Frequency of primary studies according to the consulted electronic databases. Right: Frequency of studies according to the year and the publication of the forum.

The Answer for RQ4: What are the Main Venues in Which Research in this Area has been Published? Most primary studies were published in journals. Computers & Education is the venue with the highest number of selected studies (10), followed by another journal, Computers in Human Behavior and for the proceedings of the Conference on Innovation and Technology in Computer Science Education (ITiCSE), both with six studies (Table 2).

The Answer for RQ4: What are the Main Venues in Which Research in this Area has been Published? Most primary studies were published in journals. Computers & Education is the venue with the highest number of selected studies (10), followed by another journal, Computers in Human Behavior and for the proceedings of the Conference on Innovation and Technology in Computer Science Education (ITiCSE), both with six studies.

Table 2. Main venues in which research about group formation has been published.

Name	N.
Computers & Education	10
Computers in Human Behavior	6
Proceedings of the Conf. on Innovation and Technology in Computer Science Education	6
Educational Technology & Society	5
IEEE Transactions on Learning Technologies	4
Expert Systems with Applications	3
Proceedings of the ACM International Conference on Supporting Group Work	3
Proceedings of the Conference on Computer Supported Collaborative Learning	3
British Journal of Educational Technology	2
Educational Technology Research and Development	2
Intelligent Tutoring Systems	2
User Modeling and User-Adapted Interaction	2
Total	48

4 Analysis

Drawing on the information from the primary studies, we identified that group formation usually encompasses three fundamental steps that resemble the input-process-output (IPO) model. In this context, group formation approaches receive information about students as input and then some processing or computation is performed on such information, and the outcome is information on how students should be placed in groups. These IPO steps can be performed (i) without a computational support (e.g. manually carried out by the instructor), (ii) be partially computerized (e.g. spreadsheets, LMS), or (iii) may be automatically carried out by a computer system.

Moreover, a central topic discussed in some of the primary studies is the strategy used to group students. As mentioned in Sect. 4, group formation can be carried out in a random fashion (i.e., by chance), or it can be controlled (i.e., criteria-driven group formation). As mentioned, criteria-driven group formation can fall into two categories: spontaneous group formation and controlled group formation. In the former, students are self-assigned to groups, yielding self-selected groups. In the latter, instructors or other agents handle group formation, resulting in instructor-formed groups. Evidence suggests that the main advantages of spontaneous group formation are related to self-motivation, which often leads to a more positive experience throughout the learning activities. However, since self-selected groups are made of participants that share common interests or get along with each other, the same self-selected groups might be repeatedly formed. A clear shortcoming of this approach is that some individuals might be left out. Thus, ostracized students might face difficulties coping with learning activities on their own, leading to unsatisfactory results.

Clearly, the group formation strategy influences the diversity of groups. Random approaches tend to result in heterogeneous groups. Nevertheless, since groups are formed by chance, it is possible that this approach generates single-gender groups or groups whose most elements belong to a given gender. Another problem with random approaches is that they can also form groups of students that have already mastered the subject in question or groups whose students have struggled to grasp the subject. To enforce group member diversity, therefore making the students interacting with a broad range of peers, students can be clustered in groups based on some criteria. The use of criteria adds a much more fine-grained control over group formation. Moreover, using criteria, it is also possible to create hybrid groups. As mentioned, hybrid group formation approaches employ a multitude of criteria to place students in groups that can be homogeneous and heterogeneous at the same time. However, when more criteria are employed; more information needs to be processed: this means that as the quantity of criteria increases, the need for computational support also becomes more critical. In our systematic mapping, we found that hybrid group formation is proposed along with tools that automate group formation.

Devising effective and theoretically sound group formation approaches is a notoriously complex task. Group formation approaches have to take into account many factors as for instance: (i) students' knowledge, (ii) how students interact with each other, (iii) preferences, needs or limitations of the instructors, environment and students. Given the importance of well-founded group formation approaches, we set out to investigate the rationale behind the approaches proposed in the primary studies. In a way, all primary studies justify the usefulness of their approaches. Some studies, however, fail to describe the theories, concepts, and ideas that underpin their group formation approach. This can be a worrisome finding since, according to the impossibility of justifying either theoretically or pedagogically the selection of participants to compose a group is one of the main weaknesses of the available methods, and a strong reason for teachers' hesitation in adopting systems with group formation capabilities. Finally, in order to clarify the classification scheme elaborated; and provide an example of the usage of such classification, we analyzed the group formation approach proposed by [24], as follows.

7. **Planning** – The paper aims in proposing a systematic criteria-based method for group formation that relies on genetic algorithms to process three characteristics of the students: student knowledge levels, student communicative skills, and an estimate of student leadership skills.
8. **Initiative** – The system is in charge of clustering the students based on the chosen criteria.
9. **Population diversity** – The criteria are used to form both, homogeneous and heterogeneous groups.
10. **Distribution approach** – A hybrid approach is used since, the authors main goal is to achieve groups that are as similar as possible (homogeneous) to some characteristics of the total sample of students, however also taking into account the heterogeneity of each one.
11. **Computer-supported** – Genetic algorithms are used to optimize the odds of using an arbitrary number of students' characteristics to form hybrid groups.
12. **Rationale** – The group formations approach was tested with undergrad students; however, we were not able to identify the justification (neither pedagogic nor instructional) regards the selection of the audience, the learning process, and the

learning goals. Still, the study justifies the group formation approach backing up on the benefits of using genetic algorithms to manipulate and manage a considerable number of students' characteristics. The paper comments that these three characteristics were based on a literature review.

5 Threats to Validity

A potential threat to validity is the inability to ensure that all relevant studies were selected. There is a possibility that some studies have been omitted considering that not all electronic databases were consulted. However, all consulted electronic databases are considered important forums of publications, susceptible to contain relevant studies. Thus, it can be considered that the chances that relevant studies have been resolved, or at least been mitigated. Also, we cannot eliminate threats to the validity of the quality point of view of the selected studies, because, during the selection process, we have not assigned scores to the studies. The research questions and the inclusion and exclusion criteria were drawn up before performing the systematic mapping to ensure an impartial selection process. However, the consistency of the elaborate classification scheme can also pose a threat to validity since the knowledge to elaborate it is often only achieved at the end of the selection process [23]. It is important to mention that the difference between the initial quantity of articles and the final selection can be regarded as normal in this type of study. The main objective of a systematic mapping is to provide a broad overview of which have been published in the area of interest. Due this, one important characteristic of kind of study is to avoid imposing many restrictions on the selection of primary studies [22]. Another possible threat to validity results from the operation of some electronic libraries and search systems. In some search engines, despite the use of keywords and wildcards, they often returned items fleeing the scope of the keywords used, therefore demanding further analysis of the undesired studies and manual intervention to eliminate them.

6 Concluding Remarks

This paper presented the results of our systematic mapping whose purpose was to provide an insightful overview of the current state of the art on group formation applied to CSCL. The categorization scheme we devised can be used to aid analyzing and understanding proposed group formation approaches. Moreover, it may be useful as a support tool to aid planning new ones. As a future work, we plan to optimize and extend the actual structure of the classification scheme. For example, as Massive Open Online Courses (MOOCs) are proliferating with huge number of participants with different backgrounds, culture and time zones (just to emphasize few points of interest) there is a need for new strategies for group formation in this kind of environment. Traditional approaches used for face-to-face learning or even in traditional e-Learning environments maybe cannot be appropriate for MOOCs. Although 25% of the studies state their approaches are suitable for online learning, only 1% comment that their approaches are suitable for asynchronous group learning tasks, 7% for synchronous, 19% can be used in both scenarios, however, 74% did not comment if the group interactions describe in the studies were planned to be executed synchronously or asynchronously (Fig. 7).

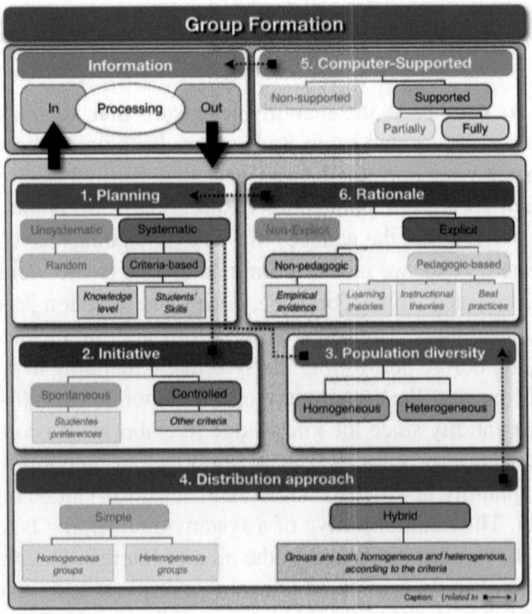

Fig. 7. Example of the usage of the classification scheme created in this systematic mapping.

Using the information extracted from the selected studies, we shed light on the most investigated approaches to group formation, the education stages in which these approaches are applied; the most used research methods, and the most common publication types. Besides the growing number of publications and the variety of computational approaches to support group formation, we identified that, the lack of presenting a rationale explaining the group formation approach can be a problem, and furthermore it can be exploited to open significant new opportunities for future research. Group formation is one of the most important steps during the planning stages of CSCL activities. Due to the many elements that need to be taken into account during group formation, devising a sound group formation approach is considered a complex task. Although research in CSCL has come a long way, most group formation efforts fail to take into account motivational aspects during group formation. Rather, we found that most existing approaches emphasize the use of basic information about the students (e.g., proficiency) during group formation. Another contribution of our systematic mapping is that by analyzing its results it is possible to identify the ways in which group formation, when applied in the context of CSCL, has been explored so far. Thus, researchers can refer to this systematic mapping to determine research gaps and decide on future research directions. Finally, we conceive an infographic that summarizes all data presented in this paper, to provide a quick overview of the results presented (Available at: http://infografico.caed-lab.com/mapping/gfc/).

References

1. Abnar, S., Orooji, F., Taghiyareh, F.: An evolutionary algorithm for forming mixed groups of learners in web based collaborative learning environments. In: Proceedings - 2012 IEEE International Conference on Technology Enhanced Education, ICTEE 2012, Kerala, pp. 1–6. IEEE (2012). https://doi.org/10.1109/ICTEE.2012.6208612
2. Adán-Coello, J.M., Tobar, C.M., de Faria, E.S.J., de Menezes, W.S., de Freitas, R.L.: Forming groups for collaborative learning of introductory computer programming based on students' programming skills and learning styles. Int. J. Inf. Commun. Technol. Educ. **7**(4), 34–46 (2011). https://doi.org/10.4018/jicte.2011100104
3. Alfonseca, E., Carro, R.M., Martín, E., Ortigosa, A., Paredes, P.: The impact of learning styles on student grouping for collaborative learning: a case study. User Model. User-Adap. Inter. **16**(3–4), 377–401 (2006). https://doi.org/10.1007/s11257-006-9012-7
4. Ardaiz-Villanueva, O., Nicuesa-Chacón, X., Brene-Artazcoz, O., de Acedo Lizarraga, M.L. S., de Acedo Baquedano, M.T.S.: Evaluation of computer tools for idea generation and team formation in project-based learning. Comput. Educ. **56**(3), 700–711 (2011). https://doi.org/10.1016/j.compedu.2010.10.012
5. Balestrini, M., Hernandez-Leo, D., Nieves, R., Blat, J.: Technology-supported orchestration matters: outperforming paper-based scripting in a jigsaw classroom. IEEE Trans. Learn. Technol. **7**(1), 17–30 (2014)
6. Barkley, E.F., et al.: Collaborative Learning Techniques: A Handbook for College Faculty. Jossey-Bass, San Francisco (2005)
7. Boticki, I., Wong, L.H., Looi, C.-K.: Designing technology for content-independent collaborative mobile learning. IEEE Trans. Learn. Technol. **6**(1), 14–24 (2013)
8. Chiu, C.-H., Hsiao, H.-F.: Group differences in computer supported collaborative learning: evidence from patterns of Taiwanese students' online communication. Comput. Educ. **54**(2), 427–435 (2010). https://doi.org/10.1016/j.compedu.2009.08.026
9. Dillenbourg, P.: Over-scripting CSCL: the risks of blending collaborative learning with instructional design. In: Kirschner, P.A. (ed.) Three worlds of CSCL: Can we support CSCL, pp. 61–91. Open Universiteit Nederland, Heerlen (2002)
10. Dillenbourg, P., Jermann, P.: Designing integrative scripts. In: Fischer, F., Kollar, I., Mandl, H., Haake, J.M., et al. (eds.) Scripting Computer-Supported Collaborative Learning, pp. 275–301. Springer, US (2007). https://doi.org/10.1007/978-0-387-36949-5_16
11. Durelli, V.H.S., et al.: A scoping study on the 25 years of research into software testing in Brazil and an outlook on the future of the area. J. Syst. Softw. **86**(4), 934–950 (2013)
12. Dybå, T., Dingsøyr, T.: Strength of evidence in systematic reviews in software engineering. In: Proceedings of the Second ACM-IEEE International Symposium on Empirical Software Engineering and Measurement (ESEM), pp. 178–187. ACM (2008)
13. Dybå, T., et al.: Evidence-based software engineering for practitioners. IEEE. Softw. **22**(1), 58–65 (2005)
14. Graf, S., Bekele, R.: Forming heterogeneous groups for intelligent collaborative learning systems with ant colony optimization. In: Ikeda, Mitsuru, Ashley, K.D., Chan, T.-W. (eds.) ITS 2006. LNCS, vol. 4053, pp. 217–226. Springer, Heidelberg (2006). https://doi.org/10.1007/11774303_22

15. Greer, J., McCalla, G., Cooke, J., Collins, J., Kumar, V., Bishop, A., Vassileva, J.: The Intelligent Helpdesk: Supporting Peer-Help in a University Course. In: Goettl, B.P., Halff, H. M., Redfield, C.L., Shute, V.J. (eds.) ITS 1998. LNCS, vol. 1452, pp. 494–503. Springer, Heidelberg (1998). https://doi.org/10.1007/3-540-68716-5_55
16. Hernández-Leo, D., Nieves, R.: Orchestration signals in the classroom: managing the Jigsaw collaborative learning flow. In: Kloos, C.D., Gillet, D., Crespo García, R.M., Wild, F., Wolpers, M. (eds.) EC-TEL 2011. LNCS, vol. 6964, pp.153–165. Springer, Heidelberg (2011). https://doi.org/10.1007/978-3-642-23985-4
17. Hsu, J., Chou, H., Hwang, W., Chou, S.: A two-dimension process in explaining learners' collaborative behaviors in CSCL. Educ. Technol. Soc. **11**(4), 66–80 (2008)
18. Ikeda, M., Go, S., Mizoguchi, R.: Opportunistic group formation. In: Proceedings of AIED Artificial Intelligence and Education, pp. 167–174 (1997)
19. Isotani, S., et al.: An ontology engineering approach to the realization of theory-driven group formation. Int. J. Comput. Collab. Learn. **4**(4), 445–478 (2009)
20. Isotani, S., Mizoguchi, R.: Adventures in the boundary between domain-independent ontologies and domain content for CSCL. In: Lovrek, I., Howlett, R.J., Jain, L.C. (eds.) KES 2008. LNCS (LNAI), vol. 5179, pp. 523–532. Springer, Heidelberg (2008). https://doi.org/10.1007/978-3-540-85567-5_65
21. Johnson, D.W., Johnson, R.T.: Learning Together and Alone: Cooperative, Competitive, and Individualistic Learning. Allyn and Bacon, Boston (1999)
22. Laurillard, D.: The pedagogical challenges to collaborative technologies. Int. J. Comput. Collab. Learn. **4**(1), 5–20 (2009)
23. Magnisalis, I., et al.: Adaptive and intelligent systems for collaborative learning support: a review of the field. IEEE Trans. Learn. Technol. **4**(1), 5–20 (2011)
24. Moreno, J., et al.: A genetic algorithm approach for group formation in collaborative learning considering multiple student characteristics. Comput. Educ. **58**(1), 560–569 (2012)
25. Muehlenbrock, M.: Formation of learning groups by using learner profiles and context information. In: Proceedings of the International Conference on Artificial Intelligence in Education, pp. 507–514 (2005)
26. Mujkanovic, A., Lowe, D.: Unsupervised learning algorithm for adaptive group formation: Collaborative learning support in remotely accessible laboratories. In: International Conference on Information Society, London, pp. 50–57 (2012)
27. Ounnas, A., Davis, H., Millard, D.: A framework for semantic group formation in education. Educ. Technol. Soc. **12**(4), 43–55 (2009)
28. Roschelle, J., Teasley, S.: The construction of shared knowledge in collaborative problem solving. In: O'Malley, C. (ed.) Computer Supported Collaborative Learning, pp. 69–97. Springer, Berlin Heidelberg (1995). https://doi.org/10.1007/978-3-642-85098-1_5
29. Stahl, G. et al.: CSCL: An Historical Perspective. University Press (2006)
30. Villasclaras-Fernández, E.D., Hernández-Leo, D., Asensio-Pérez, J.I., Dimitriadis, Y.: Incorporating assessment in a pattern-based design process for CSCL scripts. Comput. Human Behavior. **25**(5), 1028–1039 (2009)
31. Wesner, M., Pfister, H.: Group formation in computer-supported collaborative learning. In: Proceedings of the 2001 International ACM SIGGROUP Conference on Supporting GroupWork, pp. 24–31 (2001)
32. Wieringa, R., et al.: Requirements engineering paper classification and evaluation criteria: a proposal and a discussion. Requir. Eng. **11**(1), 102–107 (2005)
33. Yannibelli, V., Amandi, A.: A deterministic crowding evolutionary algorithm to form learning teams in a collaborative learning context. Expert Syst. Appl. **39**(10), 8584–8592 (2012). https://doi.org/10.1016/j.eswa.2012.01.195

Student Modelling and Knowledge Acquisition Process in Complex Nature Domains

Eleandro Maschio(✉) and Carolina Moreira

Universidade Tecnológica Federal do Paraná, Câmpus Guarapuava, Guarapuava, Paraná, Brazil
eleandrom@utfpr.edu.br, caroll-m@hotmail.com

Abstract. This chapter discusses how high-level knowledge about human expertise features can be modelled, represented and further interpreted to support learning and tutoring interactions, promoting adaptivity. For this, as an instance, the research focuses on the domain of computer programming. In the tutoring of many domains of complex nature, most past work has tended to concentrate on the theoretical principles of how humans acquire expertise. The few implementations there have been are domain-specific. However, the problem of providing an epistemology for describing knowledge about problem statements and students' states of belief has been neglected. This research treats these problems through *(1)* a method based on genetic graphs for managing the complexity of courseware authoring of problem statements and *(2)* a general process for dynamic modelling a learner's knowledge by overlaying it against the domain expertise features. Both the method and the model are supported by implemented prototype software tools that integrate the educational environment MULTIPLA. To evaluate them, empirical observations have been carried out, focusing on the generality of genetic graphs as an authoring language to provide a unified expert-student framework for developing skills.

1 Introduction

Computer programming is an introductory subject matter that influences a large portion of the Computer Science curriculum of undergraduate courses. Principles and expertise development have to fit in a limited time schedule of conceptual and practical approaches when learning such cognitively demanding tasks. The legacy of work in formal programming disciplines (both as a learner's skills and deficiencies) unleashes a whole range of ever changing problem solving techniques that rapidly includes real world situations far before graduation time.

However, according to scholarly publications in related areas, very little has been investigated as to what constitutes the specific capacities of an expert programmer's stereotype. In this sense, previous works have tended to point out the gap by stressing the sharp evolution of proficiency a learner has to face in

A. I. Cristea et al. (Eds.): HEFA 2017, CCIS 832, pp. 89–106, 2018.
https://doi.org/10.1007/978-3-319-97934-2_6

order to acquire knowledge and skill. The same literature surveys confirm this fact by raising criticism about books that claim to offer methods for developing such capacities in a few days, if not in hours [1].

Past works exposed some of the more delicate pedagogic features that have to be observed when training demands mastery level from the learners. One such example [2] is documented as the result of an investigation conducted to compare training time and skill level reached by professional musicians. Researchers found out that almost all them started at the age of 5. During the first years, training time summed up two to three hours a week. When they were 8, differences among learners started to appear. Those more inclined to reach the top positions begun to train more hours than others: six hours per week when they were 9, eight when they were 12 until they eventually made thirty hours in the beginning of their twenties. Once in there, top level musicians normally accumulate some 10 thousand hours of practice, in average. By contrast, those who did not quite fulfill the requirements of excellence accumulate 8 thousand hours and are classified as only good. Finally, there are also those who do little more than 4 thousand hours.

Thus, [1] rightly points out that deliberate practice is a key factor in expertise development for a wide variety of areas. He also suggests that carrying out tasks which are just above the current ability level of the learner should be tried and self-monitored in order to allow both performance evaluation and detection of conceptual bugs. Another study [3] claims that learning effectiveness requires strengthening of task definition by considering adequate problem difficulty to a particular person as well as informational feedback, task revisiting and error recovery scenarios.

As a consequence, we observe that ordering the statements to be presented to learners is a crucial pedagogic aspect to facilitate expertise development. We adopt statement ordering here as a general term that can be applied to any domain-specific body of knowledge. Consequently, research outcomes that embody the dynamics of a precise learner model constitutes a relevant contribution in the area of Artificial Intelligence (AI). Besides, learner modelling is not a new idea in the field of Intelligent Tutoring Systems (ITS), but past work has tended to concentrate on pedagogic issues rather than on psychological ones. This led researchers to invest more effort in proposing general frameworks that integrate architectural links from the pedagogic to the learner model and vice-versa.

Additionally, we note that little has been achieved so far in terms of concepts and tools to support intermediate-level learners in acquiring expertise. For example, on the one hand, we consider as principles of computer programming the rigid syntax and semantics of individual commands. On the other hand, expertise in the same field comprises the capacity to integrate (embed and concatenate) commands to form modular plans of high levels of abstraction. The many attempts there have been to provide automated support in learning to program divide into passive- and active-like interactive systems [4]. However, much still lies ahead in what relates to tracking a student's abilities. One way forward has been proposed under the label of multiple external representations [5], which led to considerable progress as an attempt to couple and clamp internal-to-external knowledge descriptions of tutoring systems, whether intelligent or not.

Besides putting emphasis on such compatibility, thus influencing a whole generation of tutoring systems and learning environments, the use of a multi-representational setting tends to culminate in a metacognitive approach for learners and tutors, humans or not. In theories of cognition applied to education, this usually implies in equitable division of responsibilities and improved performance of both referred actors. In a broader context, all this can be seen as the challenge of finding the best possible means of facilitating learners in their learning practice. Said otherwise, as immersive, computer-based training tasks seem to outperform traditional methods, the bottleneck of learner modelling will naturally become the next step in the AI and Education research agenda.

2 Related Work

The issues discussed here are based on three fundamental concepts: acquisition of principled and experiential knowledge; multiple external representations; problem statement sequencing from cognitive measures.

2.1 Acquiring Principles and Expertise

According to Lesgold [6], learning processes in domains of specialized and practical nature start with the acquisition of principles, followed by the development of expertise. Principled knowledge refers to the distinct bodies of knowledge which are already formalized scientifically whereas experiential knowledge involves the integration of these bodies of knowledge in real or simulated practice to produce accurate problem solutions. In fact, preceding works already pointed out a similar, dual division between specialized knowledge and knowledge acquired even with ordinary daily life.

Although a beginner's capacities concerning principled knowledge often appear to equal those of an expert's, it is exactly in experiential knowledge that many compelling differences emerge. For instance, there have been attempts to find optimal teaching/learning orderings for developing radiological expertise, given exemplar images of a class of abnormality and across different classes as well. These include ordering by typicality measures, ordering from easy to difficult exemplar cases or even ordering by combining qualitative and quantitative measures for planning tutorial instructions [6, 7]. Each problem case exposed to the learner promotes an opportunity for induction to act, altering mental categories to improve classification (almost always) a step further.

Additionally, Lesgold proposed a curriculum-based theory that focuses on expertise. In ITS, the curriculum dictates mainly content and ordering of presentation as genuine outcomes of the pedagogical decisions on how to connect students with the domain. It is still worth noticing that expertise models should not be confused with formalisms used by existing ITS architectures for curriculum definitions such as taxonomies and partonomies [8]. Expertise models are even more general formalisms. In fact, taxonomies and partonomies are just a subset among the universe of abstraction mechanisms that could be obtained by appropriately defining expertise dimensions and domain-specific teaching operations in an ITS.

According to the above text, the learning of computer programming skills can be seen as the acquisition of principles interspersed with the consolidation of expertise. As such, the principles reflect the precise comprehension of the syntax and semantics of language primitives [9]. But given the high cognitive load in computer programming, while some of the experiential knowledge does not expand enough, apprentices incur frequent errors. This poses barriers which can be very hard to overcome without appropriate instruction.

Past research works [10] have already identified the main component features of expertise. Shortly, the stereotype of an expert programmer, both in terms of principles and skills, includes items such as: (a) syntactic precision; (b) semantic precision; (c) fast pattern matching of the main source code structures based on keywords; (d) mental simulation of variable states and control flow just as symbolic execution; (e) diagnostic procedures based on a vast bug catalogue; (f) mental mechanisms for comparing equivalent program structures; (g) logic composition and assessment of command pre-conditions; (h) abstract, top-down problem analysis and decomposition; (i) concrete, bottom-up subproblem integration; (j) generalization of program modules; (k) reuse of previously developed or collected program modules; (l) solution catalogue.

However, since 1998 there has been no more research efforts that concentrated on studies of programming expertise. In the current research, we take the gap to revise existing ideas on the subject and extend them to incorporate authoring techniques to allow for formal expertise definition. In turn, future dynamic learner modelling execution will be the step forward in the research agenda.

2.2 Multiple External Representations

In the current scope, the so called External Representation (ER) [11,12] refers to the formalisms aimed at describing, organizing and presenting knowledge to humans through an interface language. The concept reminds us of a wide variety of representations, ranging from propositional, sentence-like models to strongly visual ones such as diagrams and graphics. But there are also the intermediate-level representations, which fill the gap between both extremes by associating visual and textual objects. Examples of such objects are tables, matrices, annotated scenes, parameter-based animations of physical behaviour and maps.

According to one definition [13], any meaningful ER consists of five parts: (1) the represented world; (2) the representing world; (3) a set of identifiers of the objects from the represented world being described; (4) a set of modelling aspects from the representing world; (5) a mapping function between the last two items. As a consequence of good employment of ERs in everyday life, they become a valuable resource to lower the cognitive demand of problem definition and solution during teaching and learning tasks. Besides, they apply to almost any domain of human knowledge.

Many believe that the only benefit of ERs use is to facilitate memorizing tasks by offering mnemonic objects to lower the cognitive load of certain tasks. While this is certainly a relevant feature of ERs, there are other properties that contribute to make them an important supporting formalism [12]: (1) supplying of information which can be directly perceived by the reader due to its intuitive nature (*i.e.*, no further explicit formulation is needed for human interpretation); (2) structuring of the cognitive behaviour, thus constraining the range of physical structures to be modelled (some behaviours are allowed but some others are not); (3) modifying the nature of tasks under development, allowing for easier handling of solution paths.

Besides all this, there is evidence that applying ER principles boots pedagogic approaches for problem solving in practical domains [11]. Typically, the studies suggest the adequacy of such advantageous training situations mainly based on statistical data derived from performance measures of training groups. More recently, an even more specialized research path labelled Multiple External Representations (MERs) [5] has been initiated. According to these new ideas, increase in performance measures (e.g. greater number of correct answers and shorter problem solution time) of experiments with MERs also occurred.

In the preceding studies, research effort was devoted to investigate the influence of interspersing images and text in a certain proportion. The main argument was that it would improve comprehension and lead to better performance [5]. Later, the discussions were extended to encompass a wide variety of representations during teaching sessions. Also, videos, animations and dynamic simulations became easier to produce, which led to the feedback effect of producing more because it was effective.

At the same time, researchers set themselves to question whether specific representations offered better conditions for teaching and problem solving tasks. Some others tried to assess possible problems of learning tasks involving more than one external representation, simultaneously [14]. In this latter case, the studies even support that MERs give rise to flexible ways of thinking. However, they also found that students did not benefit from the advantages of MERs possibly due to a failure to relate their different roles in concept formation.

Additionally, it can be said that generalizations concerning the efficacy of MERs in learning are somewhat difficult to be accomplished [14] because, for that, it is necessary to predict the conditions under which the chosen MERs are advantageous. Even in primary issues of problem solving, it is not yet clear how MERs can increase a learner's performance [11]. For instance, no one knows if changing the representation during reasoning about a solution is cognitively adequate.

Empirical studies [11] reported the employment of ERs in interactive learning environments aimed at aiding problem solving tasks. It was found that systems like **Bridge**, **Geometry Tutor** and **Gil** utilized graph searching algorithms as an important metacognitive resource to implement heuristic planning as pedagogic processes. In particular, **Bridge** tried to match high-level solution plans against the user's solution in a program written in Pascal. On the other hand, **Geometry Tutor** used its procedural knowledge to check the users assumptions on geometrical demonstrations. Finally, **Gil** carried out its matching procedures to tracks the user's functional organization of a Lisp program.

The same study [11] also referred to educational environments in which both artificial and human reasonings with ERs was essential. Examples of that are the **Hyperproof** system for first-order logic and the **Algebra (Word Problem Tutor)** for introductory algebraic abilities. Also stimulating reflexive thinking through the cause-effect verification approach, the **SimForest** [15] environment simulates forest growing processes. To summarize, all such systems innovated because they facilitated human learning of concepts by experimenting with glass-box and black-box mechanisms, which means that the level of interface details can be tuned by the user.

It was also found that the use of MERs principles when composing debugging resources is very important for achieving adequate visualization conditions [4]. Other evidences of the claim are also supported by parallel fields of computer science and education, such as algorithm animation and execution visualization. They all try to relate graphical (and audio) representations with textual code.

2.3 Ordering Problem Statements

Given the perspective view of expertise development, there has been effort to describe cognitive measures to support tutoring of computer programming [10]. In fact, this was part of the design of the knowledge base of problem statements of an intelligent tutoring system. Cognitive measures can work as a sturdy logical link between the domain and the learner models of the tutoring system.

Cognitive measures can play the role of qualifiers of the difficulty of problem statements and their solutions. In computer programming, such measures can be used to evaluate the learner in terms of the skills they possess when a solution is reached successfully. In this sense, the same study cited above defines the cognitive load of a problem statement as its potential to exercise the learner in writing a correct computer program. As a result, the cognitive load should also be divided in component features that are responsible for capturing the type of effect that the solution will cause on an individual learner, given his or her specific model.

In itself, the complexity of the solution program is one of the components pointed out by the study. However, since typical solution programs for beginners are quite small, the complexity can be best measured by using a mixture of lines of code (LOC) and structural complexity (SC), as well as the implementation details (ID) that are required. Structural complexity evaluates the number of loops and decision instructions that are needed, along with the degree of dependence among them to reflect their nesting situation. Implementation details refer to the number of specific preconditions that make control flow division code difficult to be written.

As well as the components concerning complexity, each problem statement influences the learner to develop expertise a step further. As a methodological issue, other dimensions of the cognitive load in a problem statement are based on measurements derived from the complexity group of features. In doing so, it is possible to quantify how much, incrementally, a problem statement demands from the learner, given his or her stereotype attributes.

So far, the picture considered here has been about the constituents of the cognitive load of a statement. However, it does not follow that each fraction of the load uniformly corresponds to a feature of expertise that the statement demands for its solution. In fact, it seems more plausible to assume that different measures possess different weights in the composition of the cognitive load of an instance problem. In turn, the weights can be elicited according to interviews with human experts in tutoring computer programming.

From a weighted equation, the cognitive measures can provide adequate quantitative means of deciding on which statement to present next. This is compatible to what human tutors already do in their everyday practice when planning human-to-human training sessions. Thus, there is relevant practical effect in classifying a statement according to its degree of contribution in face of a precise set of cognitive measures. Also useful is to create a catalogue of statements which are previously ordered by their intrinsic demands on the subset of the evolving expertise features of a typical learner.

The plausibility of the present ideas can be exemplified by the counter-situation in which a learner is presented with a statement that is substantially more complex than his or her capacity to solve. This usually leads to multiple errors, blocking the progress for long periods of time.

Under this assumption, an adverse factor in learning environments for computer programming is in the rigidity of ordering in problem statement catalogues, compared to other fields of expertise. Given fundamental aspects, even if designers of tutoring systems manage to come to a reasonable ordering of statements, there is no way of enforcing its adequacy to any learner, as real world situations include formal classes with heterogeneous background knowledge. However, it is still possible that sequencing statements can still be useful to a wide range of learners while neglecting a possibly narrow range.

Previous research contributions [16] anticipated the importance of statement complexity as a key issue in motivational aspects of pedagogic planning. In this way, proposing different statements of similar complexity may help to keep the learner motivated and productive.

Also desirable, and necessary, is the potential of tutoring environments to adapt to individual users and their competences. Few attempts have been made in this direction. One of them was the **Sequence** [10] authoring and tutoring system. Initially, its pedagogic and learner models were conceived for medical radiology but later adapted to work in the domain of computer programming. Is is based on cognitive measures of complexity (see above for LOC, SC and ID) integrated with the analysis of syntactic precision and code reuse. From that, the implemented tool is able to suggest an initial sequence of case problems, later adapting it to the user performance recorded by the learner model.

However, a wide research gap remains uncovered in the area of expertise definition and monitoring. The next step seems to be towards the dynamic association of cognitive features and complexity of problem statements in computer programming. It would also be interesting to see a new direction in investigations and developments of tutoring systems coming up as the influence of literature reviews and state of the art papers to steer the area again after a period of apparent stagnation.

3 Conceptual Framework and Software Tools

This section describes the method and software tools proposed here as a contribution to aid the development of computer programming skills. Initially, the learning environment MULTIPLA is introduced by highlighting its interactive resources, especially those related to the MERs designed for the system. After that, some of the advanced aspects of the authoring language and tool are detailed, along with example definitions of overlay learner models and problem statement sequencing.

3.1 MULTIPLA Interactive Learning Environment

The prototype software tool MULTIPLA (Multiple Languages for Typifying Interactive Learning Platforms) [9] consists of a learning environment aimed at developing computer programming skills for humans. It allows a user to acquire principles and expertise by means of three distinct interfaces: the flowchart editor; the syntax-driven coding editor; the program execution visualization mode. Figure 1 shows a snapshot of MULTIPLA while being used for relating textual Pascal (*N.B.*, other imperative languages are also dealt with here) code with its graphical appearance in a flowchart. The correspondence is built automatically and allows the user to inspect, through click based interactions, how the embedded textual structures of the Pascal program map onto the diagram lines and boxes.

The inspection mode aims at the communication of principled knowledge while the debugging mode aims at the communication of experiential knowledge. In using the inspection mode, the system-passive interface, learners search and select example programs based on problem statement information. They can also observe individual textual solution code features as well as its graphical equivalents. Moreover, they can view and change program code with standard syntax and compare complete solutions to acquire the underlying principled knowledge, necessary for understanding a program's structure and individual command structure. In using the debugging mode, the system-active interface, students stretch their experiential knowledge by engaging in execution-like dialogues where, constantly challenged by the system, they make use of their conceptual principles to form more accurate and complete mental models of program behaviour.

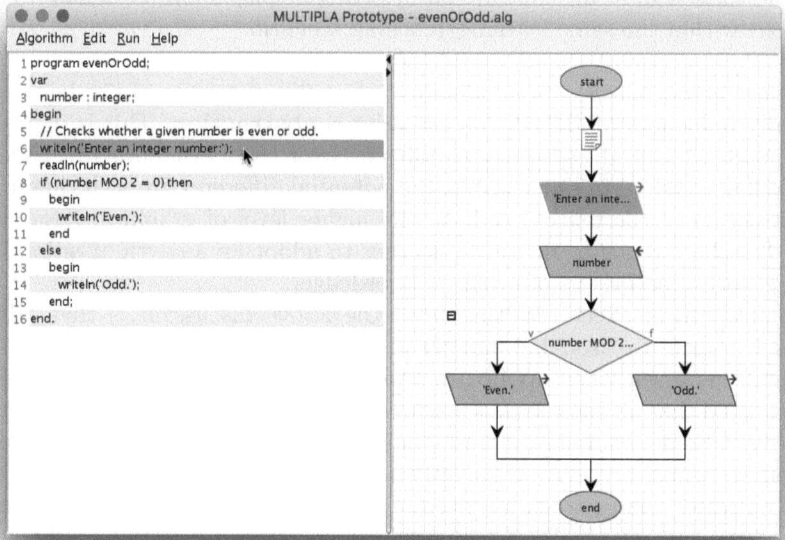

Fig. 1. Snapshot of MULTIPLA's interface

The debugging mode offers two ways of execution: stepwise and timed. Both reveal the exact stage of execution by showing its variable values and by highlighting its textual command on the code as well as its corresponding box on the flowchart. By synchronizing the highlights, MULTIPLA manages to externalize one of the most delicate types of dynamic behaviour in the form of visual knowledge. Related to that, one might say that physical phenomena are also hard to express through computer metaphors. However, the pragmatic fact is that designers of learning systems usually find quite neat forms to externalize physical knowledge, for instance, of kinematics. By contrast, designers of systems for computer programming face barriers to forge metaphors of linguistic aspects, which could be briefly summarized here as a paradoxical attempt to find concrete world object to represent what is abstract in its own nature (*e.g.*, a grammar). For this, multiple representations arise as a plausible option to rely on when investing original research effort.

The multi-representational setting introduced here brings up similar requirements to those of the DeFT (Design, Functions,Tasks) conceptual framework [5] aimed at the teaching of various domains with the support of MERs. Thus, a parallel between MULTIPLA and DeFT is presented next, focusing on some of the most relevant semantic resources underlying Ainsworth's proposal of functional taxonomy MERs [5]. As a matter of clarity, it is important to note that classifications based on the referred taxonomy are non-excluding ones, meaning that the same ER in an educational system can play more than one role (or function) within the same learning/teaching scenario.

(1) Constructing Deeper Understanding. For this category, the subclass that markedly influences MULTIPLA's ER roles is **abstraction**. This is due to the correspondence that the system maintains between the textual program code and its flowchart version. In other words, the potential difference from the lower level of abstraction of the textual code to the higher level of graphical information of the flowchart tends to invite the learner to adopt an analytic position when relating both descriptions of program knowledge.

An interesting effect resulting from the above discussion is that, after a learner has seen a reasonable number of example programs performing similar computations (even when applied to apparently different problem statements), it is expected that all the relevant features related to the program will be learned inductively. Some example programs might even display very little in common with others which are more frequently seen (*e.g.*, searching and decision problems), but after a large stock has been inspected, the learner is expected to have acquired the "prototypical" view of each problem class in question. Although we do not yet make use of explicit statistical knowledge, this is not in disagreement with more traditional theories of concept formation [17] which states that an idealized or abstracted view of a problem class is based on a combination of concepts assimilated from many typical (or frequent) features of instances.

The other two role subclasses here concern **relation** and **extension**. The former also refers to the relationship between the textual code and the flowcharts while the latter includes the dynamic interpretation of the code through the debugging mode. This is done by highlighting the program control flow steps as well as the variables valuations.

(2) Constraining Interpretation. Under this role class of Ainsworth's taxonomy, the relevant function here to the MULTIPLA environment is the one of constraining by *inherent properties*. In this sense, even when learners are not familiar with any of the ERs presented by the system interface, the most specific one (*e.g.*, a textual code in Pascal) can help to constrain the ambiguities of the more generic one (*e.g.*, a flowchart, since its graphical information encompasses the grammar of a wide variety of textual languages). It is worth observing that covering the abstract-to-concrete range of functions is mainly different from covering the generic-to-specific one, although both forms of analysis and classification bear similarities.

(3) Complementing Roles. The approach adopted by MULTIPLA under this class of function is mainly one of showing **different information**, rather than different processes. This is the case since the automatic interpretation of program structures is the same for both the textual and graphical forms. However, the information contained in them may play an uncountable number of effects on learner, as it has been reported by so many previous publications.

The colours adopted in MULTIPLA were also designed to play complementary roles. One of the roles is in relation to the dynamic effect of the debugging mode, especially that of the memory cells valuation changes. The other one is used for grouping semantically similar flowchart objects (boxes and lines) and stressing their relationship with the textual code during inspections by clicking.

To evaluate the educational effectiveness of MULTIPLA, an experiment has been carried out in the domain of introductory programming. The positive results obtained from the evaluation tend to indicate, among other findings, that MULTIPLA is an appropriate environment of MERs to teach about domains of programming expertise. A full description of the experiment can be found in an earlier work [9].

Additionally, to substantiate generality claims about MULTIPLA, we are starting to combine the educational evaluation of the leaning environment with the definition of learner models and problem statement sequencing for domains of programming expertise. But in order to do that, we have just finished extending MULTIPLA with authoring tools that allow these definitions. The next subsection describes such tools.

3.2 Learner Modelling and Statement Sequencing

As described in Sect. 2, problem statement sequencing is a crucial aspect when instructing learners on how to conduct their educational journey. This might include different ways of ordering, such as by typicality measures, from easy to difficult ones, or even ordering by combining qualitative and quantitative measures for planning instruction. In this sense, three statements are listed below, according to their cognitive load, from the least to the most complex (*i.e.*, generally, from easy to difficult).

Statement 1
Given the age of a person as input, write out if he has reached the age of majority.

Statement 2
Given the age of a person as input, write out if voting is obligatory. The obligation to vote holds if the citizen is between 18 and 70 years old.

Statement 3
Given the age of a person as input, write out if voting is obligatory, optional or forbidden. The obligation to vote holds if the citizen is between 18 and 70 years old. Voting is optional between 16 and 18, or above 70.

Now, assuming that a learner has solved the second problem, it would be interesting and quite plausible to provide intelligent mechanisms in a tutoring system (*e.g.*, an ITS) to choose Statement 3 as the next challenge. But, depending on the learner's performance during the solution of Statement 2, it might also be a good option to backtrack to Statement 1. Such mechanisms are realizable if the tutoring system can count on a precise model that dynamically adjusts the learner profile in accordance with knowledge of the domain. Besides, it is necessary that a human tutor (author) manually classifies each statement according to the fraction of domain knowledge that it tends to demand for being solved.

The present work has adopted the *overlay* model of the learner, also called genetic graph [18]. The approach consists in representing the learner knowledge as a subset of the domain knowledge. Thus, by successively overlaying both sets, a tutoring system can infer (or replan) pedagogic objectives. And, obviously, this assumes that domain and learner models are codified under compatible formalisms.

Initially, we studied and formulated the basis of a formal language to capture the features of expertise in computer programming to allow for the overlaying process to be carried out automatically. Also, the correlation among various features was observed, since the study pointed this out as an important statement sequencing factor. As the underlying formalism of our general framework, the genetic graph consists of facts and rules (vertices of a graph) connected by relationships (edges) that evolve mainly in following ways: (a) from simplified to more elaborate concepts; (b) from error to correctness; (c) from abstract to refined details of command sequence; (d) from specialized to general subroutines; (e) from pre- to post-requisite code conditions.

Figure 2 illustrates a simple, stylized diagram to represent a seven-expertise learner model (the dark subgraph) overlaying many-concept domain model (the pale graph, including the dark part). The matching process of the overlaying mechanism is not yet implemented since the focus of the current stage has been on the authoring tools. However, the heuristic search algorithms that will support the matching process are specified and partially implemented.

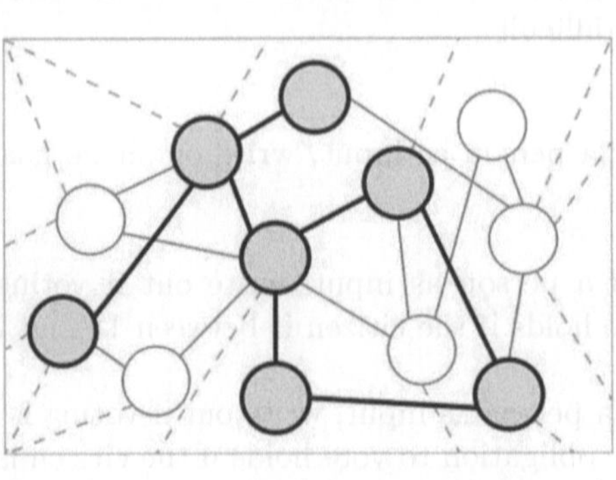

Fig. 2. Learner model overlaying domain model

At present, we are also specifying and implementing the differential process that will output a set of recommended concepts and problem statements to teach next. Thus, assuming that there are still features of expertise in the current learner profile, the search algorithm might be able to find it and return a list of priority tasks. Additionally, we have employed the same representation and matching process to the problem statements. Since each statement demands specific expertise features to be solved, it can also be seen as a genetic subgraph of the domain.

Figure 3 shows a stylized diagram to represent a two-expertise subpart of the domain model (the darkest subgraph), which in theory, will be computed from the overlaying process.

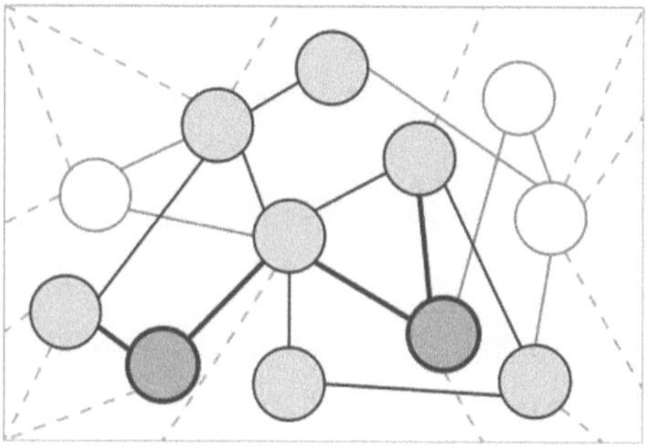

Fig. 3. Output subgraph of the method

Figure 4 shows four stylized diagrams of the catalogue problem statements demanding different expertise features as authored by a human tutor along with their correct answer.

Given the above explanations, the learner model gains it dynamic nature through an interpreter that is fed with the solutions given by users of the tutoring system. Currently, the evaluation and feeding process is still manually done by a human tutor. We are endeavouring to replace the human tutor by a bug diagnostic module, but that has proved one of the hardest bits to be automated in MULTIPLA.

Another possibility guaranteed by the standard genetic graph representation consists in learner group evaluation. From that, the server system could provide the human tutor with a mirror of the typical expertise features under development. This could be useful, for instance, even for real life tutoring lessons since neglected points of the curriculum would be dealt with accordingly.

But limitations in the current implementation still pose certain difficulties to model gradual changes in individual expertise features. At present, an expertise feature is either possessed by the learner or not. MULTIPLA has no power to reason with confidence factors or other probabilistic data. However, we are also studying ways of empowering the system in agreement with more recent theories of concept formation. We are also planning to provide rich information to human tutors about the temporal stage of development in which a learner is found in a given moment. In other words, a history of expertise features can be provided by the system for a specific period of the course. Finally, we stress that modelling through overlaying offers an invaluable setting for supporting pedagogic tasks.

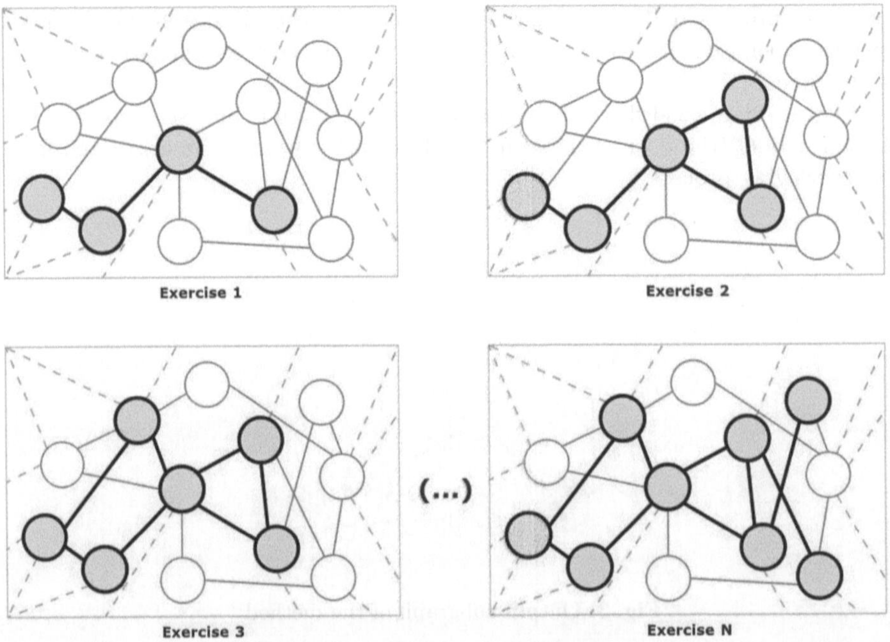

Fig. 4. Four problem statement graphs

Figure 5 shows sequenced stylized diagrams of a learner model (dark subgraphs) in a time line. Relating to this, we also planning to use pale and strong colours to display the strength of an expertise feature when we are able to compute fuzzy probabilities to it.

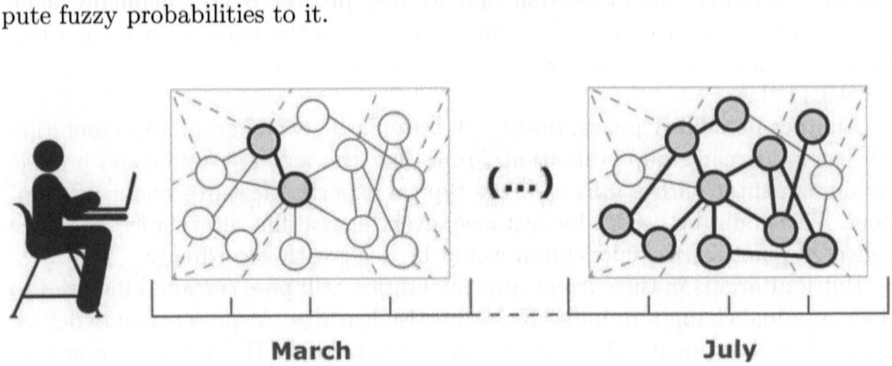

Fig. 5. Evolving learner model in a time line

Finally, we stress that modelling through overlaying offers an invaluable setting for supporting pedagogic tasks. None of the ideas presented in the subsection, whether already implemented or not, are too complex to achieve (except for automatic bug diagnosis). However, all of them are useful to guide human tutors and learners in their routine.

3.3 Authoring Tools

The authoring process is supported by two implemented prototype software tools. The first is aimed at the domain model description in which the details of feature expertise are provided to the system. Figure 6 shows a snapshot of the tool interface while being used to edit a node of the genetic graph of expertise. It allows the WYSIWYG manipulation of the graph elements. A number of options are allowed: (a) inserting, deleting and editing of expertise features; (b) establishing relationships among expertise features; (c) defining initial properties of the genetic graph; (d) validating the graph structure by running reachability of any feature. The output from this tool is filed in XML format.

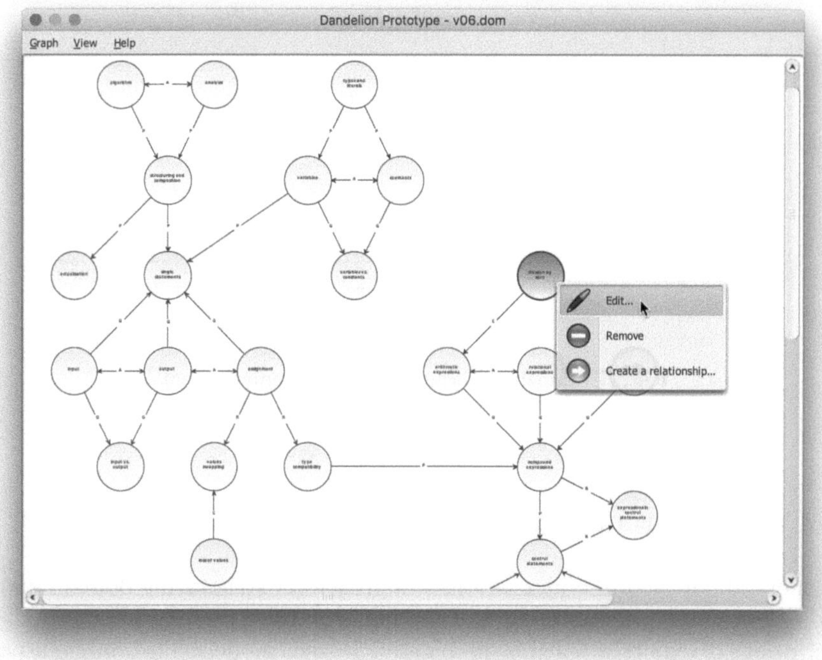

Fig. 6. Snapshot of the domain model authoring tool

The second tool is aimed at the elicitation of problem statements which are described as subgraphs of the domain model. Figure 7 shows a snapshot of the tool interface while being use for defining a specific problem statement. This is done through expertise features selection to inform which ones are demanded for the solution to be accomplished correctly. The tool also allows inspection on which statements overlap their demands and which expertise features are unattended by at least one statement.

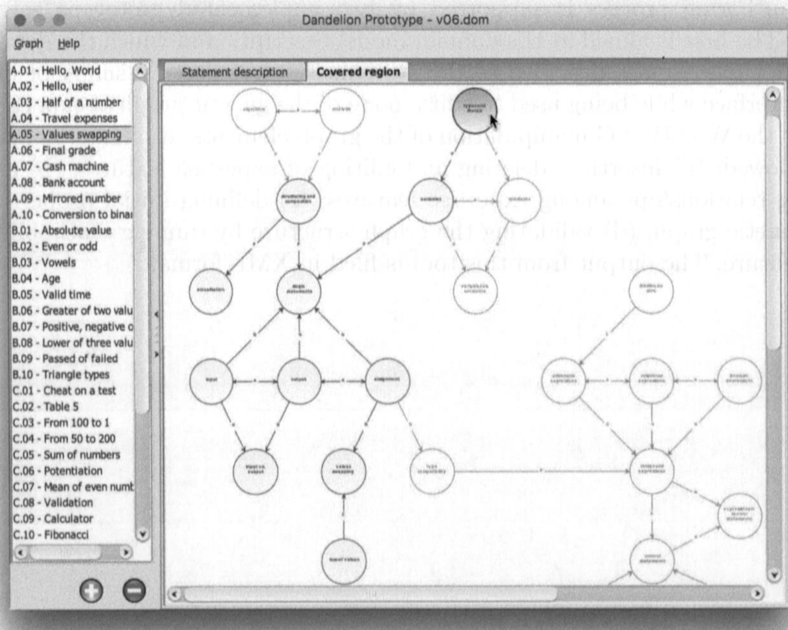

Fig. 7. Snapshot of the problem statements tool

4 Conclusion and Future Research

The method outlined provides a formal approach for structuring domain knowledge of computer programming and mechanisms for reasoning about symbolic descriptions of expertise features in dynamic learner models. Theoretical aspects of the method and mechanisms have been discussed, highlighting solutions for specific situations like the use of genetic graphs [18] for overlaying learner models against domain models and ordering of problem statements in tutorial sessions. The authoring and learning/tutoring prototype tools, called MULTIPLA, are in an advanced stage of implementation and are capable of allowing authors to go through a cycle for developing courseware material for computer programming.

To substantiate generality claims about MULTIPLA, we are combining the evaluation of the software tools with the definition of knowledge bases for domains of expertise in the imperative programming paradigm, such as in typical Pascal and C programming. The evaluation procedures have been concentrating on the power of the authoring formalisms as well as on the learning effectiveness of the tool. The positive results obtained so far suggest that multiple external representations Ainsworth [5] and genetic graphs form a suitable pedagogic framework in the context of ITS systems.

Finally, despite the substantial results already achieved with the first version of MULTIPLA, many points can be improved in the authoring tools as well as in the learning/teaching mechanisms. The near boundary of future research concentrates in evaluating and improving the authoring potentialities of the framework. The primary attempted is to test two hypotheses: (a) the genetic graph is general enough, as a language, to produce a wide range of learning and tutoring environments for computer programming, based on expertise features; (b) the interactive tools that support the authoring process are appropriate in the sense that they not only accelerate the process but also provide an unified framework for developing skills.

Acknowledgements. The authors would like to thank the Guarapuava Campus of the Federal University of Technology, Paraná (UTFPR), for the financial support through the Public Notice 01/2017 – DIRPPG-GP.

References

1. Norvig, P.: Teach yourself programming in ten years (2001). http://norvig.com/21-days.html
2. Gladwell, M.: Outliers: The Story of Success, 1st edn. Little, Brown and Company, New York (2008)
3. Lave, J.: Cognition in Practice: Mind, Mathematics and Culture in Everyday Life. Cambridge University Press, New Cambridge (1988)
4. Romero, P., Cox, R., du Boulay, B., Lutz, R.: A survey of external representations employed in object-oriented programming environments. J. Vis. Lang. Comput. **14**(5), 387–419 (2003)
5. Ainsworth, S.: DeFT: a conceptual framework for considering learning with multiple representations. Learn. Instr. **16**(3), 183–198 (2006)
6. Lesgold, A., Rubinson, H., Feltovitch, P., Glaser, R., Klopfer, D., Wang, Y.: Expertise in a complex skill: diagnosing x-ray pictures. In: The Nature of Expertise. Lawrence Erlbaum Associates Inc., Hillsdale (1988) 311–342
7. Direne, A.I., Scott, D.: Identifying the component features of expertise in domains of complex visual recognition. Information Technology Research Institute Technical report Series, University of Brighton (2001)
8. Howard, R.W. (ed.): Concepts and Schemata. Cassel Educational, London (1990)
9. Maschio, E., Direne, A.I.: Multiple external representations to support knowledge acquisition in computer programming. Braz. J. Comput. Educ. (RBIE) **23**(3), 81–96 (2015)
10. Pimentel, A.R., Direne, A.I.: Medidas cognitivas no ensino de programação de computadores com Sistemas Tutores Inteligentes. In: Anais do IX Simpósio Brasileiro de Informática na Educação, SBIE 1998, Fortaleza, CE, Brasil, vol. 20, pp. 205–215. Sociedade Brasileira de Computação (1998)

11. Cox, R., Brna, P.: Supporting the use of external representations in problem solving: the need for flexible learning environments. J. Artif. Intell. Educ. **6**(2/3), 239–302 (1995)
12. Zhang, J.: The nature of external representations in problem solving. Cogn. Sci. **21**(2), 179–217 (1997)
13. Palmer, S.: Fundamental aspects of cognitive representation. In: Cognition and Categorization, pp. 259–303. LEA, Hillsdale (1978)
14. Ainsworth, S., Labeke, N.V.: Using a multi-representational design framework to develop and evaluate a dynamic simulation environment. In: Dynamic Information and Visualisation Workshop, July 2002
15. Murray, T., Winship, L., Bellin, R., Cornell, M.: Toward glass box educational simulations: reifying models for inspection and design. In: External Representations in AIED: Multiple Forms and Multiple Roles - AI-ED 2001 Workshop, San Antonio, Texas, EUA, IOS Press, May 2001
16. Soldado, T.D., du Boulay, B.: Implementation of motivational tactics in tutoring systems. J. Artif. Intell. Educ. **6**(4), 337–378 (1995)
17. Mervis, C.B., Rosch, E.: Categorisation and natural objects. Annu. Rev. Psychol. **32**, 89–115 (1981)
18. Goldstein, I.P.: The genetic graph: a representation for the evolution of procedural knowledge. Int. J. Man Mach. Stud. **11**(1), 51–77 (1979)

Coupling Cultural Context to a User Model of Adaptive E-Learning: Fundamentals, Approach and Experiment

Isabela Gasparini[1(✉)], Marcelo S. Pimenta[2], José Palazzo M. de Oliveira[2], and Ricardo Miranda[3]

[1] Universidade do Estado de Santa Catarina (UDESC), Joinville, Brazil
isabela.gasparini@udesc.br
[2] Universidade Federal do Rio Grande do Sul (UFRGS), Porto Alegre, Brazil
{mpimenta,palazzo}@inf.ufrgs.br
[3] Universidade Federal do Paraná (UFPR), Curitiba, Brazil
ricardo.miranda@nsn.com

Abstract. Considering culture both in technology-enhanced learning and human computer interaction research is an important issue since culture has a strong impact on many cognitive and affective processes including those related to learning. Also, people with different cultural backgrounds develop alternative interpretations and strategies and do not value their environment in the same way, and this reflects on their interactions and satisfaction with learning technologies, especially e-learning systems (ELSs). Thus, adapting ELS to specific users, taking advantage of knowledge acquired about their cultural, technological, educational and personal characteristics is an essential feature in order to improve usability and flexibility of ELS. This paper presents the context-aware and culture-oriented aspects of an adaptability approach designed for ELSs and validated by an experiment conducted with a group of 65 students of three different countries to evaluate their satisfaction in the culture-oriented adaptation provided by the system. Based on the data of the participants, the results suggest that the culturally adapted interface has more participant satisfaction than the non-cultural one.

Keywords: Cultural context · User model · Adaptive e-learning

1 Introduction

Although normally used by different types of users, with different profiles and cultures, an e-learning system (ELS) is usually created and maintained without taking into account a diversity-orientation. In times when web-based systems need to pro-vide support for an ever increasing amount of material and make it available for local-language populations across the world, the introduction of the culture concept in web-based systems is becoming a necessity, a challenge, and a timely and relevant issue. In fact, considering culture in technology-enhanced learning is an important issue since culture has a strong impact on many cognitive and affective processes related to learning. Also, people with different cultural backgrounds develop alternative interpretations and strategies consequently reflecting in their interactions with learning

A. I. Cristea et al. (Eds.): HEFA 2017, CCIS 832, pp. 107–123, 2018.
https://doi.org/10.1007/978-3-319-97934-2_7

technologies. Thus, it is not a surprise that culture has received increasing attention in the Adaptive E-Learning community.

Student modeling is one of the major research issues in adaptive and intelligent educational system fields. Typical existing student models contain information about students' interests; knowledge, background and skills; interaction preferences; individual traits and sometimes the students' learning styles. However, ELSs may be dynamically adjusted not only according to the students' model but also depending on a richer notion of context (Eyharabide et al. 2009). Although there are several approaches to consider context in ELS, they focus mainly on technological and networking aspects without regarding other contextual aspects such as cultural and educational profiles.

This paper presents a new approach to user modeling in adaptive e-learning, taking into account cultural context complementary to 3 other types of contextual information, technological, educational and personal. The cultural context aims to improve personalization capabilities, making use of a set of ontologies to represent cultural properties explicitly as an extension of traditional student modeling. A user's cultural context shapes his/her perception of system features: a given culture context will possibly cause a user to focus on a set of information and ignore others. Thus, system features appropriate for one culture may not be suitable for others; and system design needs to be easily adapted to different cultures as well. As the testbed of this research, an actual open-source ELS called AdaptWeb® (Adaptive Web-based learning Environment), currently in use by some Brazilian universities, was extended in order to incorporate our approach of cultural-aware context-based adaptation.

The main features of our approach are described and illustrated, showing how to use it to model context and culture for personalization of ELS AdaptWeb®, representing the rich context explicitly as an extension of traditional student modeling. In an effort to evaluate our ideas, an experiment with a group of 65 students of three different countries was conducted, focusing on some usability aspects of AdaptWeb® extensions and the users' satisfaction in the culture-oriented adaptation provided. Both performance (effectiveness and efficiency) and preferences (satisfaction) of the target population were then measured. The paper also describes how the experiment was planned, conducted and the results further interpreted.

2 Literature Review

Human Computer Interaction (HCI) researchers agree that the design of technologies requires culture-based considerations (Young 2011) and previous research indicates that usability can only be assured if future systems are designed in a culture-oriented way (Reinecke and Bernstein 2007). The influences of culture can be seen in products, by choices concerning use of colors, symbols, language set, and so on, and in the design process, e.g. culture influences higher level design issues, design methods employed in building interfaces and usability methods. The cultural influence, if ignored, can compromise usability evaluations, therefore giving information that is inaccurate (Vatrapu and Pérez-Quiñones 2006).

For example, in the learning domain, Guo and Reinecke (2014) present an empirical study of how students navigate through Massive Open Online Courses. They investigated how navigation strategies differ by demographics such as students' age and country. People from different cultures use user-interfaces in different ways, expect different graphical layouts, and have different expectations and patterns in behavior. Therefore, user-interfaces have to be adapted to the needs of different locales, i.e. they are then called culturally adapted. Culture is a term used by social field researchers to refer to a set of parameters that can be used to differentiate a social group significantly, such as nations, companies and groups (Abou-Zeid 2005), and cultural differences have been studied by many anthropologists (e.g. Hofstede, Hall, Trompenaars and Hampden-Turner). Cultural adaptation, in practice, means modifying the actual content, design and style of a system and addresses culture to user's requirements. It requires cultural awareness of the design and respective language, customs and the specific characteristics of the corresponding target users.

It is possible to generalize what users consider usable and attractive in some extent, but, some aspects, are certainly a matter of personal taste, or strongly influenced by cultural values (Reinecke and Bernstein 2011). The same conventions apply to user interface preferences, which highly vary across cultures, e.g. Asian web sites are commonly bright and colorful, and their high complexity is often perceived as information overload by Westerners, who prefer more structured content. Users also differ in their design preferences and in their perception of usability, so, according to Reinecke and Bernstein (2011), it's important to consider such culturally-determined partialities for certain Look & Feel of user interfaces.

Different researches on the implications of cultural aspects in the user interface design were carried out, and several elements that can be adapted based on Hofstede's national dimensions were discovered (Hofstede 2001), (Hofstede 2005). The adaptations rules in interaction design are summarized for each Hofstede dimension, showing the recommendations to low and high scores of each dimension. Table 1 shows the main adaptation rules, based on Reinecke and Bernstein (2011); Reinecke and Bernstein (2013); Marcus and Gould (2012); Marcus and Gould (2000a, b), Reinecke et al. (2010), Marcus (2006), Callahan (2005), Wan Mohd Isa et al. (2007), Barber and Badre (1998), Marcus (2002). Research conducted on the effect and usability of culturally adapted web sites and interfaces has already shown enormous improvements in working efficiency (Reinecke and Bernstein 2007). Adaptation process is an initiated and oriented system, and it can dynamically adapt the content, presentation, interaction design, and the assistance offered to users, according to the user model.

Table 1. Adaptation rules versus Hofstede's Dimensions

Hofstede	Adaptation for high and low scores	
	Low score	High score
PDI	Non-linear navigation preferred	Users expect to get instructions, high level of support and many navigational cues
	Different access and navigation possibilities	Linear navigation, few links, minimize navigation possibilities
	High information density, most information at first sight	Low information density
	Hierarchy of information not deep, with most information at interface level	Little information at first level
	Friendly error messages suggesting how to proceed	Rigorous error message

(continued)

Table 1. (*continued*)

Hofstede	Adaptation for high and low scores	
	Low score	High score
	Asymmetry, shallow hierarchies	Symmetric, deep hierarchies
	Many options of functionalities	Reduce choice of functionalities
	Less highly structured	More highly structured, guide access to information
IDV	Colorful interfaces preferred	Monotonous colors for interface
	High multi-modality	Low multi-modality
	Possible use of Multimodal interfaces	High text to image ratio
	Group images, Emphasis on the state of "being"	Images of individuals, emphasis on action
	Individual roles downplayed (e.g. just product); group focus Preference for socially supportive and constrained claims Controversy discourage: tends to divide people	Focus on maximizing personal achievement Images of young/activity, rather than age/wisdom/ "being"
MAS	Non-linear navigation preferred; exploration and different paths to navigate	Navigation oriented towards exploration and control; Restrict navigation possibilities
	Not much saturation; use pastel colors	Bright colors, with high contrast
	Multiple choices	Limited choices
	Orientation towards relationships	Orientation towards goals
	Attention gained by visual aesthetics	Graphics used for utilitarian purposes
	Gender/work roles blurred Mutual exchange and support more important than master Website task-oriented and provides quick results for limited tasks More emotional/aesthetic appeal	Traditional gender/family/age distinctions emphasized; work tasks/roles preferred Mastery most important; websites designed for exploration, control Games/competitions grab attention Artwork may be utilitarian/instrumental
UAI	Complex interfaces, most information at interface level	Organize information hierarchically
	Non-linear navigation, with exploration permitted	Linear navigation paths, show position of the user
	Variety of choices; long pages with scrolling	Limited choices; limited scrolling
	Maximize information with code colors, sounds and typography	Redundant cues to reduce ambiguity
	Less controlled navigation: links may open windows, change site Help system focus on information; task orientation secondary	Keep it simple Reveal results/implications of actions Make attempt to prevent looping/becoming "lost"
LTO	Little information density	High density information, most at interface level, menus with few levels
	Highly structured content, small units	Focal area to arrange the content
	Presentation of information with high hierarchy	Less structure data
	Little information density	High density information, most at interface level, menus with few levels

Typically, the most widely used components of student profiles have been: knowledge, individual traits such as learning or cognitive styles, experiences and background, goals or tasks (Brusilovsky and Millán 2007). In recent years, researchers have also added contextual information.

In a broader sense, context describes the circumstances under which something occurs as well as the interrelationships of those circumstances. Such interrelationships provide a semantic perspective that restricts and narrows the meaning of "something" (Abarca et al. 2006). The term context has been interpreted in different ways in different domain areas and diverse approaches have been applied to capture and use of the contextual information (Jovanovic et al. 2009). A context-aware ELS is an application that adapts its behavior according to the students' context. Context-aware applications not only use context information to react to a user's request, but also take the initiative as a result of context reasoning activities (Dockhorn Costa et al. 2007). Thus, an improvement in the user's contextual information leads to a better understanding of users' behaviors in order to adapt (i) the content, (ii) the interface, and (iii) the assistance offered to users.

Research in adaptive educational hypermedia has demonstrated that considering context leads to a better understanding and personalization (Brusilovsky and Millan 2007). On the one hand, context is vital to improve personalization/adaptation in ELSs. Recent works aim to provide the capacity for identifying the right contents, right services in the right place at the right time and in the right form based on the current student's situation (Barbosa et al. 2006), (Rosa et al. 2005), (Bouzeghoub and Do Ngoc 2008). The interesting propositions of GlobalEdu (Barbosa et al. 2006), (Rosa et al. 2005), in terms of architecture, for instance, have distributed and central alternatives with different models (student, context and environment). Jovanović et al. (2009) developed the LOCO (Learning Object Context Ontologies), an ontological framework to allow formal representation of the notion of learning context, that is composed by a specific learning situation, determined by the learning activity, the learning content and the student involved. A work related with situations and scenarios is the research presented in Yang (2006), where the author proposes to model adaptive and context-aware scenarios based on: a didactics theory; a domain model; a learner model and a context model. To develop the scenario, several scenarios based on a common learning scenario is first built; secondly, a theory in didactic anthropology of knowledge to acquire the scenario model and the didactical environment is used; after, the didactic based scenario model is transformed into a computer based hierarchical task model. Different of our work, Yang doesn't work with the automatic description of a user situation, but with the description of adaptive scenarios.

The student's cultural aspects refer to preferences and ways of behavior patterns determined by the person's culture. According to Guzman and Motz (2005) cultural characteristics can be described on different levels, such as national and regional, organizational, professional, and individual characteristics and may be related to several different kinds of culture, such as national, ethnic, religious, community of practice/interest, among others. Enriching ELSs with cultural differences can help understand the students' behavior even better, and enable personalized learning according to cultural needs. Particularly about cultural aspects, Blanchard and Mizoguchi (2008) describe an upper ontology of culture. They use this approach to deal with many CATS (Culturally-Aware Tutoring Systems) related issues by providing objective formalism for cultural representation. Chandramouli et al. (2008) presented the notion of the CAE-L Ontology for modeling stereotype cultural artifacts in adaptive education and used a Cultural Artifacts in Education questionnaire to gather the information required to determine if there is a significant cultural bias within online education, specifically Adaptive Educational Hypermedia. Mohammed and Mohan (2013) present the idea of contextual student modeling, referring to the process of building a computational representation of the cultural identity and background of a student. Reinecke et al. (2007) present a Cultural User Model Ontology - CUMO that contains information such as different places of residence, the parents' nationality, languages spoken, and religion. CUMO contains information about Hofstede's five dimensions and their values: (a) MAS - Masculinity vs. Femininity, (b) UAI - Uncertainty Avoidance Index, (c) PDI - Power Distance, (d) IDV - Individualism vs. Collectivism and (e) LTO - Long Term Orientation. CUMO also takes into account all places of residence and calculate their influence on the user's dimensions according to the duration of the user's stay at those places. Our research has a different point of view as we integrate cultural, technological, educational and personal aspects as part of a rich context model (complementary to student model). Despite the considerable scientific production on customization, personalization and adaptation in e-learning, surprisingly there is very little research work devoted to methods to incorporate context-aware and culture-oriented concerns to ELSs.

3 Approach to Modeling Context and Culture

We have developed a model based on upper-level ontology. In this model, a student might be involved in several overlapping contexts, and consequently, his/her educational activity might be influenced by the interactions between these contexts. Overlapping contexts contribute to and influence the interactions and experiences that people have when performing certain activities (Bouzeghoub and Do Ngoc 2008), (Eyharabide and Amandi 2008). As described in Eyharabide and Amandi (2008), our model has three levels of abstraction: meta-model, model (ontologies) and object. The meta-model level is represented by an upper ontology, describing abstract concepts like user, application, situation or date; the model level expresses the different contextual dimensions, with several ontologies to describe the elements that populate the context and, in the lowest level - the object model, we find the instantiations of the context ontologies, e.g. a concrete name of a specific user (e.g. Ada Lovelace), a specific discipline (Software Engineering), or a particular device (mobile device). In other words, the ontology concepts of one level are the instantiations of its immediate superior level. We personalize an ELS for each user based on the information stored in a student model. In our work, the typical characteristics of students are extended to include the context dimensions having personal, technological, educational and cultural aspects. *Personal context* includes the student's personal information and preferences. The characteristics of the user in our model include: age, gender, background, navigation mode and help assistance, and some information about the domain, including the student's course information (e.g. Computer Science, Mathematics, etc.), and student's discipline (e.g. Artificial Intelligence, Database system, etc.). *Educational context* consists of multifaceted knowledge due to many distinct viewpoints of pedagogical information needed to personalize e-learning. In practice, our Educational Ontology considers the users' knowledge of the subject being taught and Felder's learning styles (Felder and Brent 2005). *Technological context* is related to the different technological constraints, and we are concerned nowadays especially about these elements: screen resolution, type of device (computer or mobile device), operating system, browser type and version, speed navigation, download rate, IP and localization (city, state and country), internet provider, and the network connection (bandwidth). *Cultural context* includes the cultural background of a student and may have a great impact on his/her ability and efficiency to learn a given set of content (Chandramouli et al. 2008). Cultural context refers to different languages, values, norms, gender, social, ideological, political or ethnic aspects (Pawlowski 2008). Our Cultural Ontology uses Hofstede's national dimensions, the student's mother tongue, other foreign languages and skills, educational background and countries where the student has lived (exchange programs), and the student's location (IP address) because people can be in a different country of their birth, and then, we can show them, besides learning objects related to their culture, some aspects of the culture they are involved in at that moment. In the future, we want to extend this model to incorporate other cultural issues and dimensions (e.g. emotional, personal, social, organizational) and other models, e.g. Hall, Trompenaars and Hampden-Turner. However, we first plan to measure student's performance and satisfaction in other experiments with different students groups.

Figure 1 presents the proposed model. The meta-model is an upper-level ontology describing abstract concepts like user, application, user profile, situation or date. The model depicts the different contextual dimensions. Each contextual dimension is represented by a different ontology such as a cultural ontology, education ontology (course, discipline, etc.), personal ontology (gender, preferences, birthday, etc.) or technological ontology (operating system, browser, network bandwidth, etc.). Finally, the object model will comprise instances describing the context of a particular user like a concrete name (John Smith), a course (Human Computer Interaction) or a particular language (English). The situation could be completely changed if the contexts of students change. Among all the possible information gathered in the student model, we are especially interested in modeling scenarios because they change according to context. Scenarios may depend on the situation the student is now in and on external factors. The concreteness of scenarios helps students and teachers to develop a shared understanding of the proposed contextual information, and allows assimilating and representing complex idiosyncrasies of what they would otherwise misunderstand.

We define a scenario as a tuple containing an entity that the student prefers in a given situation, a relevance denoting the student's preference for that entity, a certainty representing how sure we are about the student having that preference and a date to indicate when that preference is stored.

$$Scenario = \{entity, situation, relevance, certainty, date\}$$

Situations are the key to include temporal aspects of context in a comprehensive ontology for context modeling, since they can be related to suitable notions of time

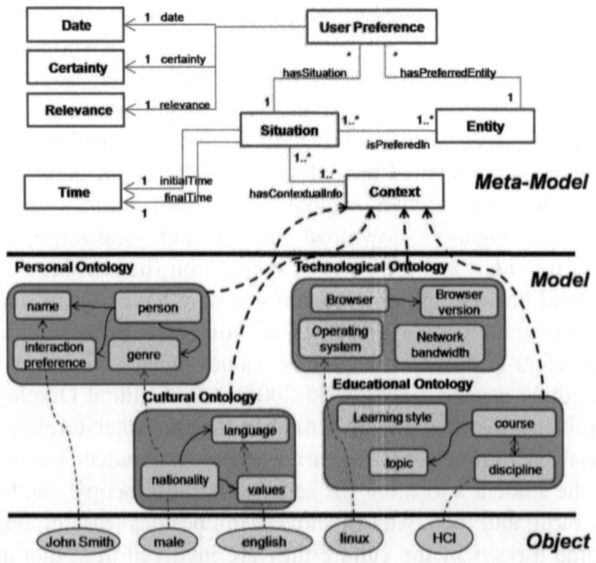

Fig. 1. Example of a scenario-oriented situation

(Dockhorn Costa et al. 2006). As context varies during certain time intervals, it is vital to consider it within the concept of Situation. Examples of situations could be "John was at home using his notebook to read lesson number 3 of the Human Computer Interaction course" or "A Japanese Professor, who speaks English, is adding new exercises to the course Introduction to Java using a high speed connection while she travels by train". Therefore, we define situation as a set of contextual information in a particular period of time.

$$Situation = \{Context, initial\ time, final\ time\}$$

An example of contextual information would be: "The student named John is reading lesson number 7". This is a description relating an entity (the student John) to another entity (the lesson number 7) via a property (is reading). We represent this contextual information as (Student.john, isReading, Lesson.lesson#7). We define the context as a set of triples composed by concepts, instances and relations between them. It is important to emphasize that the concepts and instances might belong to the same ontology or different context ontologies.

$$Context = \{(Ca1.Ia1, R1, Cb1.Ib1), \ldots, (CaN.IaN, RN, CbN.IbN)\};$$
$$(C\!: concept, I\!: instance\ and\ R\!: relation)$$

Let us consider the John example. John prefers reading visual learning material in a situation when he is at home using his notebook to read lesson number 3 of the Human Computer Interaction course. Hence, the corresponding context1 will be:

Context1 = {(Person.John, locatedIn, Location.home), (Person.John, uses, Device.notebook), (Person.john, reads, Lesson.lesson#3), (Lesson.lesson#3, belongsTo, Course.HCI)}
Situation1 = {Context1, 5:00PM, 8:00PM}
Scenario1 = {User, Situation1, relevance.high, certainty.95%, date.02-02-2014}

4 Architecture for Context-Aware ELS

The extended AdaptWeb® architecture is presented in Fig. 2. Starting with the modules, the User Interface Component is responsible to both obtain the user data, and present the adaptations processed by the environment. Actually, the AdaptWeb® environment already stores all the information related with the login, the chosen discipline and the author notifications to the students. So, it is possible to aggregate user context data to be obtained via interface, like the learning object actually in use and the path made by the student while using AdaptWeb®. Knowing this path, we can discover the occurrence of learning events that are important to start an adaptation. These events are detected by the Context Collector/Detector and, depending on the event, notified to the Context Management Service.

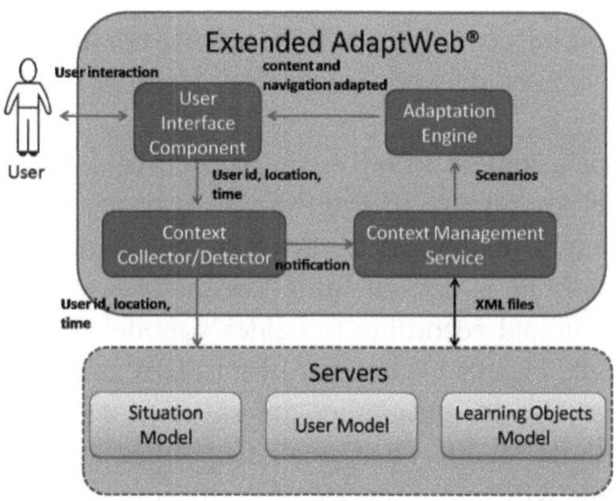

Fig. 2. Extended architecture of AdaptWeb®

The extended architecture is based on three servers that operate together to provide and manage contextualized data according to the student's scenarios. Each server manages specific data related to the user context, being respectively responsible for the storage and adaptation of (i) environmental context (information related to the user environment, tasks, activities, time interval, devices, location), (ii) information about students (personal data, preferences, objectives, knowledge background, behavior, learning styles, cultural context, etc.), and (iii) learning object information (documents provided by the educational environment to its users for their learning).

The Context Management Service is responsible for analyzing the context managed by the servers, generating different scenarios that can be experienced by the students in a specific period. These scenarios are used to guide the adaptation (in the Adaptation Engine), and materialized in the interface rendered to the user. The main goals of the architecture are: (i) easily reuse educational resources, since they will be adapted to the user scenario while the stored content remains the same, (ii) integration into the existing architecture, since the new architecture is supposed to take advantage of the existing functionalities and (iii) extensibility to other educational systems, using standard technologies. The personalization is possible with the combination of contextual data related to whom and where the user is, what he/she is doing and what he/she needs to achieve his/her educational targets.

4.1 Adopting Context Modeling

In this section, we explain how the adaptation process of the system was improved by student's cultural, educational, personal and technological contexts, in order to support the recommendation of learning objects and to provide the best materials to the students. For each scenario, the system is dynamically personalized depending on the context information available. We show some examples in a Database Systems course context where the teacher provided a set of links of learning objects with diverse content about database system, for example: History and motivation for database systems, Components, DBMS functions, Database architecture and data independence, etc. For a simplification purpose, we just have a few variables of the student's model: student's knowledge, subject, network connection, learning style, Language, Language Level and Country.

In *Context1*, João is a student who lives in Brazil, his mother tongue is Portuguese, and he has a low level knowledge of English. He is trying to learn about the subject XML databases, which is explained in English. He is doing exercises about that subject, but unfortunately he is not obtaining satisfactory results. In addition, he has a high network connection and according to Felder's model (Felder and Brent 2005) he is active. The user model checks his number of mistakes and identifies if he needs help resolving the exercises. In the meantime, the situation model detects, via teacher's agenda, that a chat with the students was previously scheduled by the teacher to happen in 15 min. These events will start a service of notification in the Context Management Service, informing that a change of the current scenarios related with these events may change. After a new orchestration by the Context Management Service, the User Interface sends a message to the student, notifying him of this possibility to solve his doubts and shows the "chat" link in a different and highlighted color.

In another scenario, *Context2*, Maria, a Spanish speaking PhD student of Engineering from Argentina, has very good skills in three different foreign languages (English, Portuguese and French). She is also learning the subject XML databases and not having good results. She has a low network connection and her Felder's learning style is reflective. In consequence, AdaptWeb® sends a message by email to her teacher advising to contact the student and changes the order of the links, putting links related to video material with low quality resolution in the end and disabling links related to video material with high quality resolution (those which are heavy and difficult to see).

Furthermore, AdaptWeb® detects some important links for learning material written in English and French and shows this in the top of the list.

These contexts are formalized as following:

Context1 = {
(Student.*João*, **isLearning**, Subject. *XML databases*),
(Subject. *XML databases*, **isExplainedIn**, Language.*english*),
(Student. *João*, **hasUserKnowledge**, UserKnowledge.*bad*),
(Student. *João*, **hasConnection**, NetworkConnection.*high*),
(Student. *João*, **hasStyle**, LearningStyle.*active*),
(Student. *João*, **hasMotherTongue**, Language.*portuguese*),
(Student. *João*, **hasLanguageSkill**, Language.*english*),
(Student. *João*, **hasEnglishLanguageLevel**, LanguageLevel.*low*),
(Student. *João*, **isCitizenOf**, Country.*Brazil*)}

Context2 = {
(Student.*Maria*, **isLearning**, Subject. XML *databases*),
(Student. *Maria*, **hasUserKnowledge**, UserKnowledge.*bad*),
(Student. *Maria*, **hasConnection**, NetworkConnection.*low*),
(Student. *Maria*, **hasStyle**, LearningStyle.*reflective*),
(Student. *Maria*, **hasMotherTongue**, Language.*spanish*),
(Student. *Maria*, **hasLanguageSkill**, Language.*english*),
(Student. *Maria*, **hasEnglishLanguageLevel**, LanguageLevel.*high*),
(Student. *Marie*, **hasLanguageSkill**, Language.*portuguese*),
(Student. *Marie*, **hasPortugueseLanguageLevel**, LanguageLevel.*high*),
(Student. *Marie*, **hasLanguageSkill**, Language. *french*),
(Student. *Marie*, **hasSpanishLanguageLevel**, LanguageLevel.*high*),
(Student. *Marie*, **isCitizenOf**, Country.*Argentina*)}

In summary, the adaptation mechanisms in AdaptWeb® can be the following actions/recommendations, for example:

Context1 → "send notification to student only in Portuguese" + "show highlighted links" + recommend LO and content about the same subject (same concept in the domain ontology) in Portuguese with a low level of difficulties.

Context2 → "order links" + "hide or disable links" + "show highlighted links" + "recommend LO and content about the same subject written in Spanish, English, French or Portuguese".

There are two main categories in HCI to support adaptation and localization of the interface (Heimgärtner 2007), (i) presentation of information, e.g. time, date, color format, language, font, writing direction, etc.; and (ii) dialog design, e.g. menu structure and complexity, layout, positions, navigation concept, interaction path, interaction speed, system structure, etc. Based on the literature review, we summarize the adaptation rules in interaction design for each Hofstede dimension, for Argentinean, Brazilian and Uruguayan users.

5 Experiment

This research is focused on examining the value of cultural adaptations to system usability, especially on users' satisfaction - by the following research question: Is system usability improved by cultural adaptations? In the current study, we seek answers to the above question by empirically examining the user subjective satisfaction of the cultural adapted system. Our hypothesis is:

H0: The culturally adapted design has no effect on user subjective satisfaction.
H_1: The culturally adapted design has an effect on user subjective satisfaction.

The goal of the experimental study was to evaluate the subjective satisfaction of the culturally adapted system, especially in term of its interface, navigation, content and assistance personalized for the user. We evaluated the participant's perceived satisfaction with the interaction experience, and the data was collected through a questionnaire.

5.1 Participants

First we conducted two pilot tests with ten students to verify our method of gathering information and of evaluating the system. In our experimental study, regarding participants, there were 65 students of the computer science field, from Argentina, Brazil and Uruguay, recruited by partner universities in an international cooperation project. The prerequisites to participate were (i) to have some knowledge of database and SQL (Structured Query Language), (ii) to be unfamiliar with the environment and (iii) to accept to be part of the research (term of consent). We selected the SQL teaching topic because of its technical nature, trying to eliminate some bias of cultural aspects present in some contents and disciplines of humanities fields (e.g. policy, law, etc.).

5.2 Procedure

We divided students in different groups, randomized. Each group had to perform six tasks related to SQL, three in the current environment (not adapted to the cultural aspects – 'Nad', namely 'Discovery' for the participants) and three in the cultural adapted environment ('Ad' – namely 'Atlantis' for the participants). The order of the utilization of each environment and the selection of the tasks were randomized as well, to avoid a possible impact of learning effects. The aliases ('Atlantis' and 'Discovery') were used so that people did not identify the system.

All steps of the experimental study were completed in a month. Fist we invited the students to participate of the experiment. Then, they had to register, by filling personal, cultural and pedagogical data. After that, they had to use it in another discipline, prepared in order to give the student some experience in the system. Only students who had taken all the previous steps were invited to continue participating in the experiment day. The students were asked to do six tasks and then, to answer a questionnaire about their satisfaction. Also, in parallel we observed the user interaction with the system and we stored all data in log files. All students had the same resource conditions (i.e. standardized software, a desktop platform, access to Internet) on the experiment day. The SQL materials (learning objects, examples, exercises and complementary materials) were available in English, Portuguese and Spanish and they were mainly texts and images.

5.3 Data Analysis and Results

Of the 65 participants in the experiment, 87.69% were male and 12.31% female. Dividing by each country, Argentina had 93.33% males and 6.67% females; in Brazil 92.50% were males and 7.50% were female, and in Uruguay 60% were male and 40% were female. Unfortunately the low rate of women in the experiment is a reflection of female rates in computer science classes in general. It is not our intention to discuss this issue here or even how to change it. On the age of the participants, in Argentina all participants were between 18–25 years; in Uruguay 90% were between 18–25 years and 10% between 34–41 years, in Brazil 85% of the participants were in the 18 to 25 year range, 12.5% were between 26–33 years and 2.5% in the 34–41 year range. Compared to other languages, the profile raised in Argentina was that 93.33% had knowledge of other languages (all participants had some knowledge of English language, one participant knew Italian and another knew Japanese). In Uruguay, 90% of the respondents knew other languages. Of these, everyone knew English, 22.22% knew French and 11.11%, Portuguese. In Brazil, 50% of respondents knew other languages. Of these, 60% knew English, 5% spoke Spanish, 20% English and Spanish; 5% English, Spanish and Italian, 5% English and French, 5% English, Spanish and Portuguese. The Portuguese language appeared in the results because we had one participant from Germany who is living in Brazil. Of all the participants from the three countries, two Brazilians had lived in other countries.

A usability questionnaire was prepared and validated in two pilot experiments, in order to evaluate students' satisfaction. The questionnaire had 42 questions, where: 31 questions were divided in eight groups of topics based of UI elements: (a) Navigation; (b) Hierarchy; (c) Information density; (d) Image and icons; (e) Structure; (f) Color and Color saturation; (g) Guidance and (h) Usability. Each topic was evaluated regarding both interfaces (adapted to cultural issues- 'Ad' and not adapted- 'Nad') with a 5-points Likert scale; five questions were about the cultural adaptations provided by the system in terms of content, presentation and navigation– with a 5-points Likert scale; and six open questions were about positive and negative aspects of both interfaces.

In order to analyze the first set of questions (the 31 questions about both interfaces, split in eight groups), we first applied the Shapiro-Wilks test of normality. For each group, the test results indicated data did not have normal distribution. Since the compared variables were discrete ordinal, we applied the Wilcoxon test to evaluate whether the difference between the medians was zero, *i.e.*, if there is no difference between the interfaces. Some tests performed are listed below, and they all showed a significant difference in the medians of each group, *i.e.*, the sample provides evidence that there is significant difference between the cultural adapted interface ('Ad') and the not adapted one ('Nad').

Table 2. Results

Aspect Evaluated	Adapted interface median score	Not adapted median score	p-value
Navigation aspect	4	2	3.49e−09
Hierarchy	4	2	1.665e−11
Information Density	4	3	7.653e−12
Images	5	2	3.977e−12
Structure	4	2	3.465e−12
Color and Color saturation	4	2	3.77e−12
Guidance	4	3	4.443e−12
Usability	4	2	1.005e−11

As we can see in Table 2, all eight groups of topics had the null hypothesis rejected, *i.e.*, the median values of each interface had a significant difference. In all groups, the culturally adapted interface had a better performance. These were the first analysis, and further investigation must be carried out to reach/obtain a deeper understanding. About the questions related to the cultural adaptations provided by the system in terms of the content, presentation and navigation, the participants highlighted the importance of receiving materials in other languages in which they have some proficiency - the majority of participants (80%) agree (partly agree and totally agree), indicating that the vast majority of participants verifies the importance of receiving examples and complementary materials in other languages; 10.77% participants neither agree or disagree and 9.23% disagree (partly or totally). One of the open questions was about the participants' preference of interface and navigation and why they preferred it. Of the 65 students, 78.46% preferred the 'Atlantis' interface, *i.e.*, the cultural adapted interface and navigation; 18.46% preferred the 'Discovery' interface, *i.e.*, the regular interface, without the cultural adaptations; 1.54% did not answered and 1.54% liked both interfaces equally. About the other goals of usability, i.e., effectiveness (in terms of tasks performed by the participants) and efficiency (in terms of number of steps taken by the participants to accomplish the tasks between the two environments), we have found no statistical evidence.

6 Conclusions

Culturally-aware systems are those that use cultural aspects to influence or to provide information and/or services relevant to a task execution. Designing a culturally-aware system is not a trivial task, since it is necessary to deal with issues associated to: which kind of information should be considered as a cultural aspect, how to represent this information, how to assess the most appropriate representation structures to model culture and to provide culture-awareness, how it can be acquired and processed and how to design the culture usage into the application. Thus, culture-awareness is nowadays an interesting research area and receives increasing attention in both technology-enhanced learning and Human Computer Interaction community, particularly in intelligent and adaptive system fields.

In this paper, we presented an approach to user modeling in e-learning, taking into account (among other rich contextual information) the cultural context to improve personalization capabilities. In the current version, our system context information is based on Hofstede's five national dimensions, the student's mother tongue, other foreign languages and skills, educational background and countries where the user lived (exchange programs), and the student's location (IP address). Some of these cultural properties are obtained automatically (for instance, student location) while others need some user effort with form filling or questionnaire answering. After acquired, this information is used to adapt content, navigation and presentation of the system. We conducted an experimental study with 65 students from Argentina, Brazil and Uruguay, comparing the current system and the culturally adapted one. The results indicate the cultural interface is perceived more satisfactorily by the participants. More experiments and analysis must be done, in order to deeper investigate the cultural aspects.

We offer our process as one potential way of conducting experimental design studies, which should not be taken as set in stone but as a starting point for a vivid and creative discussion about alternative approaches and ideas. Our process focuses on design studies as conducted by HCI researchers. Interesting questions rising from this focus include: how does the process generalize to practitioners? How does it differ when applied to other learning situations? We hope that our work will entice more HCI researchers into this fast-growing part of the field, and that it will inspire further methodological discussion. Our future work will investigate more complex characteristics of cultural context in the adaptation process of our environment.

References

Abarca, M.G., Alarcon, R.A., Barria, R., Fuller, D.: Context-based e-Learning composition and adaptation. In: Meersman, R., Tari, Z., Herrero, P. (eds.) OTM 2006 Part II. LNCS, vol. 4278, pp. 1976–1985. Springer, Heidelberg (2006). https://doi.org/10.1007/11915072_106

Abou-Zeid, El-S: A culturally aware model of inter-organizational knowledge transfer. Knowl. Manag. Res. Pract. **3**, 146–155 (2005)

Barber, W., Badre, A.: Culturability: the merging of culture and usability. In: Proceedings of the 4th Human Factors and the Web Conference (1998)

Barbosa, D.N.F., Augustin, I., Barbosa, J.L.V.: Proceedings of the Fourth Annual IEEE International Conference on Pervasive Computing and Communications – Workshops (2006)

Blanchard, E.G., Mizoguchi, R.: Designing culturally-aware tutoring systems: towards an upper ontology of culture. In: Workshop on Culturally-Aware Tutoring Systems (CATS)-ITS (2008)

Bouzeghoub, A., Do Ngoc, K.: A situation based meta-data for describing pervasive learning objects. In: mLearn: 1st International Conference on Mobile Learning (2008)

Brusilovsky, P., Millan, E.: User models for adaptive hypermedia and adaptive educational systems. Adapt. Web **4321**, 3–53 (2007)

Callahan, E.: Cultural similarities and differences in the design of university websites. J. Comput.-Mediat. Commun. **11**(1), 12 (2005)

Chandramouli, K., Stewart, C., Brailsford, T., Izquierdo, E.: CAE-L: an ontology modelling cultural behaviour in adaptive education. In: 3th International Workshop on Semantic Media Adaptation and Personalization, pp. 183– 188. IEEE Computer Society (2008)

Dockhorn Costa, P., Almeida, J.P.A., Ferreira Pires, L., van Sinderen, M.: Situation specification and realization in rule-based context-aware applications. In: Indulska, J., Raymond, K. (eds.) DAIS 2007. LNCS, vol. 4531, pp. 32–47. Springer, Heidelberg (2007). https://doi.org/10.1007/978-3-540-72883-2_3

Dockhorn Costa, P., Guizzardi, G., Almeida, J., Pires, L., Van Sinderen, M.: Situations in conceptual modeling of context. In: 10th IEEE on International Enterprise Distributed Object Computing Conference (2006)

Eyharabide, V., Gasparini, I., Schiaffino, S., Pimenta, M., Amandi, A.: Personalized e-Learning environments: considering students' contexts. In: Tatnall, A., Jones, A. (eds.) WCCE 2009. IAICT, vol. 302, pp. 48–57. Springer, Heidelberg (2009). https://doi.org/10.1007/978-3-642-03115-1_5

Eyharabide, V., Amandi, A.: Semantic spam filtering from personalized ontologies. JWE – J. Web Eng. Rinton Press **7**(2), 158–176 (2008)

Felder, R., Brent, R.: Understanding student differences. J. Eng. Educ. **94**(1), 57–72 (2005)

Guo, P.J., Reinecke, K.: Demographic differences in how students navigate through MOOCs. In: First ACM Conference on Learning@ Scale Conference, pp. 21–30. ACM (2014)

Guzman, J., Motz, R.: Towards an adaptive cultural e-learning system. In: LA-WEB - 3th Latin American Web Congress. IEEE Computer Society (2005)

Heimgärtner, R.: Towards cultural adaptability in driver information and assistance systems. In: Aykin, N. (ed.) UI-HCII 2007 Part II. LNCS, vol. 4560, pp. 372–381. Springer, Heidelberg (2007). https://doi.org/10.1007/978-3-540-73289-1_43

Hofstede, G.: Cultures and Organizations: Software for the Mind, 2nd edn. McGraw-Hill, New York (2005)

Hofstede, G.: Culture's Consequences: Comparing Values, Behaviors, Institutions and Organizations Across Nations, 2nd edn. SAGE Publications, Thousand Oaks (2001)

Wan Mohd Isa, W.A.R., Md Noor, N.L., Mehad, S.: Incorporating the cultural dimensions into the theoretical framework of website information architecture. In: Aykin, N. (ed.) UI-HCII 2007 Part I. LNCS, vol. 4559, pp. 212–221. Springer, Heidelberg (2007). https://doi.org/10.1007/978-3-540-73287-7_27

Jovanović, J., Gasevic, D., Torniai, C., Bateman, S., Hatala, M.: The social semantic web in intelligent learning environments - state of the art and future challenges. Interact. Learn. Environ. **17**(4), 273–308 (2009)

Marcus, A.: Cross-cultural user-experience design. In: Barker-Plummer, D., Cox, R., Swoboda, N. (eds.) Diagrams 2006. LNCS (LNAI), vol. 4045, pp. 16–24. Springer, Heidelberg (2006). https://doi.org/10.1007/11783183_4

Marcus, A.: Dare we define user-interface design? Interactions **9**(5), 19–24 (2002)

Marcus, A., Gould, E.W.: Globalization, localization, and cross-cultural user-interface design. In: Jacko, J.A. (ed.) Human–Computer Interaction Handbook Fundamentals, Evolving Technologies, and Emerging Applications, Chapter 15, 3th edn, pp. 341–366. CRC Press (2012)

Marcus, A., Gould, E.W.: Crosscurrents: cultural dimensions and global web user-interface design. ACM Interact. 7(4), 32–46 (2000a)

Marcus, A., Gould, E.W.: Cultural dimensions and global web user-interface design: what? So what? Now what? In: 6th Conference on Human Factors and the Web, Austin, Texas (2000b)

Mohammed, P., Mohan, P.: Contextualised student modelling for enculturated systems. In: AIED 2013 Workshops, Proceedings of the 4th International Workshop on Culturally-Aware Tutoring Systems (CATS 2013), vol. 5, pp. 20–29 (2013)

Pawlowski, J.M.: Culture profiles: facilitating global learning and knowledge sharing. In: The 16th ICCE, pp. 537–544 (2008)

Reinecke, K., Bernstein, A.: Knowing what a user likes - a design science approach to interfaces that automatically adapt to culture. MIS Q. **37**(2), 427–454 (2013)

Reinecke, K., Bernstein, A.: Improving performance, perceived usability, and aesthetics with culturally adaptive user interfaces. ACM Trans. Comput.-Hum. Interact. **18**(2) (2011)

Reinecke, K., Bernstein, A.: Culturally adaptive software: moving beyond internationalization. In: Aykin, N. (ed.) UI-HCII 2007 Part II. LNCS, vol. 4560, pp. 201–210. Springer, Heidelberg (2007). https://doi.org/10.1007/978-3-540-73289-1_25

Reinecke, K., Schenkel, S., Bernstein, A.: Modeling a user's culture. In: The Handbook of Research in Culturally-Aware Information Technology: Perspectives and Models. IGI Global (2010)

Reinecke, K., Reif, G., Bernstein, A.: Cultural user modeling with CUMO: an approach to overcome the personalization bootstrapping problem. In: Workshop on Cultural Heritage on the Semantic Web, ISWC (2007)

Rosa, G.P.J, Ogata, H., Yano, Y.: A multi-model approach for supporting the personalization of ubiquitous learning applications. In: IEEE International Workshop on Wireless and Mobile Technologies in Education, pp. 40–44 (2005)
Vatrapu, R., Pérez-Quiñones, M.A.: Culture and usability evaluation: the effects of culture in structure interviews. J. Usability Stud. **1**(4), 156–170 (2006)
Yang, S.J.H.: Context aware ubiquitous learning environments for peer-to-peer collaborative learning. Educ. Technol. Soc. **9**(1), 188–201 (2006)
Young, P.A.: The significance of the Culture Based Model in designing culturally aware tutoring systems. AI Soc. **26**(1), 35–47 (2011)

Improving the Metacognitive Ability of Knowledge Monitoring in Computer Learning Systems

Tiago Roberto Kautzmann(✉) and Patrícia Augustin Jaques

Universidade do Vale do Rio dos Sinos (UNISINOS), São Leopoldo, Brazil
tkautzmann@gmail.com, pjaques@unisinos.br

Abstract. Knowledge monitoring is a fundamental metacognitive skill for the development of other metacognitive skills and is related to successful learning in all academic settings, including higher education. This skill allows students to identify their knowledge gaps more precisely, so they can engage in self-regulated activities. Self-regulated learners are proactive in setting goals, planning and deploying study strategies, monitoring the effectiveness of their actions and taking control of their learning and more prone to achievement in college. The knowledge monitoring skill becomes increasingly more important for basic education students who will enter higher education, given the demands for higher thinking skills in college. This chapter aims to present an Animated Pedagogical Agent (APA) that trains the metacognitive ability of knowledge monitoring in computer learning systems. The training provided by the APA encourages learners to reflect on their knowledge and has its content and frequency of intervention adapted to the characteristics of each student. Results from evaluations with seventh and eighth grade students supported the hypothesis that students who receive the training improve both their knowledge monitoring skill and their performance in the domain. Results also suggest a less hasty behavior from students who received the APA training. These findings indicate that the proposed APA can help students in basic education to acquire essential metacognitive skills to enter higher education.

Keywords: Knowledge monitoring skill · Metacognition
Adaptive instruction · Animated pedagogical agent · Computer learning systems

1 Introduction

The last few decades have seen an increase in the use of metacognition processes to improve students' learning in computer learning systems [3, 4, 10, 11, 15, 17, 18, 25, 28]. Such works investigate how to improve students' skills in monitoring their own cognitive processes and the impact of metacognition in learning. A subgroup of these studies investigates the more specific skills of self-regulated learning [3, 15, 17]. Self-regulated learners are proactive in setting goals, planning and deploying study strategies, monitoring the effectiveness of their actions and taking control of their learning [29]. Self-regulated learners are assumed to be metacognitive learners, since self-regulated learning can only exist through metacognitive processes such as cognitive monitoring and control [6].

A. I. Cristea et al. (Eds.): HEFA 2017, CCIS 832, pp. 124–140, 2018.
https://doi.org/10.1007/978-3-319-97934-2_8

Schoenfeld [20] has demonstrated the importance of metacognition in learning. One of the findings presented by the author is that learners who are more efficient in problem solving spend more time reflecting on the problems than less efficient students. Schoenfeld observed that even students who had enough Math skills to solve the problems potentially presented inefficient behavior for not spending enough time reflecting on the problems. Inefficient students in the experiment did not take the opportunities they had to stop what they were doing and reflect on whether their strategy was working or if they should try a new strategy. These students spent a considerable amount of time trying to solve the problems, but little time reflecting on the problems.

Monitoring cognitive processes is a fundamental part of metacognitive processes, since it allows individuals to infer the metacognitive knowledge required for cognition control [6]. Metacognitive knowledge is a person's knowledge or beliefs about factors or variables that affect the progress and the results of their cognitive endeavors [8]. The knowledge a person has about not having enough knowledge to solve a problem is an example of metacognitive knowledge. It allows people to make efforts to acquire deficit knowledge. Studies by Tobias and Everson, discussed in [23, 24], found that the knowledge monitoring skill is fundamental in the acquisition of other metacognitive skills. The metacognitive skill of knowledge monitoring is the ability people have of identifying what they know and what they do not know about a given subject. Students who are able to accurately identify their knowledge in a subject are prone to work harder in the development of their deficit areas [9], to seek help when necessary [21] and to study more strategically [23, 24]. Students are less likely to engage in more sophisticated metacognitive activities such as planning, deploying study strategies and learning assessment, when they fail to identify what they know and what they do not know about a certain domain [23].

The work in [23] presents significant relationships between college and high school grades and knowledge monitoring ability. Tobias and Everson [7] also found that knowledge monitoring skill scores also predicted first year college grades. Learning in complex domains, such as those found in higher education, often requires that students recall substantial amount of prior knowledge to understand and acquire new knowledge or solve problems. Some prior knowledge may be recalled imperfectly, or may never have been completely mastered. Students who can accurately distinguish between what they know and do not know are more likely to review and to relearn imperfectly mastered materials, compared with those who are less accurate in estimating their own knowledge [7]. The transition from basic education to college puts many demands on students. The greatest challenge may be the move from the declarative knowledge emphasized in basic education to the higher level thinking typically required in college. Students may need a variety of self-regulated learning and metacognitive skills that were not necessarily essential in high school [12]. The knowledge monitoring skill is fundamental for students to acquire other essential skills for college, such as self-regulated learning skills [23], so it may be worth teaching basic education students to improve their metacognitive ability of knowledge monitoring to acquire the college prerequisites of higher thinking skills.

The primordial goal of our study is to identify whether the knowledge monitoring skill can be improved with training. Students can be taught to improve their metacognition [5, 14], and that is true even in computer learning systems [2]. However, previous work on training exclusively the knowledge monitoring skill in computer learning systems was not found. We assume that reflecting on problems should be the basis of metacognitive training, since previous work has shown that this reflection is essential in metacognitive processes such as cognitive monitoring [22]. None of the related studies we found adapts the portion of the instruction that encourages student's reflection on personal knowledge [1, 3, 10, 18, 28].

The present work aims at investigating the specific effects of knowledge monitoring training in computer learning systems that adapt system content and frequency of intervention to the students' metacognitive skill level, task performance, and problem solving history. The proposed training encourages learners to reflect on their knowledge during problem solving processes. An animated pedagogical agent (APA) that trains the metacognitive skill of knowledge monitoring was developed to achieve that goal. This chapter is an extended version of our previous work, presented in [13]. In this chapter, we deepen the description of the metacognitive agent proposed and also expand the data and analysis of the results obtained in the experiment. We also present a discussion about the importance of training the knowledge monitoring skill of students beginning higher education.

For evaluations purpose, the APA was integrated into PAT2Math, an algebraic step-based Intelligent Tutoring System (ITS). ITSs have intelligent modules that infer students' knowledge among other types of information. Accordingly, integrating the training agent into an ITS allows us to implement the adaptive characteristics of the proposed metacognitive training. Additionally, step-based ITSs allow agents to perform training intervention in each step of the problem solving process.

The evaluation of the training proposed had 107 seventh and eighth grade students from private schools in São Leopoldo, Brazil. The experiment followed a pretest-posttest design with control and experimental groups and it aimed to answer the following research questions: (1) Does the adaptive training provided by the agent improve knowledge monitoring?; (2) Does the adaptive training provided by the agent improve performance? The success of this study would present some evidence for the training of an essential ability for students to enter higher education.

This chapter is organized as follows. Section 2 presents related works. Section 3 presents the animated pedagogical agent that trains the knowledge monitoring skill. This section describes how the agent assesses the student's current level of knowledge monitoring, how it works, its implementation, and its integration into the ITS. The method is described in Sect. 4, whereas results are shown in Sect. 5. Section 6 discusses findings, and conclusions are drawn in Sect. 7.

2 Related Works

The related works we present below describe computer learning systems that at some point of the instruction use messages or other types of visual aid to encourage students to reflect on their knowledge.

Gama [10] proposed the reflection assistant model, which aims to train knowledge monitoring and other two metacognitive skills. The training in this model encourages students to reflect on their knowledge, but all activities in the model occur in a fixed sequence, i.e. they are not adapted to specific characteristics of the students. The findings of this study were inconclusive, so that it was not possible to identify a clear positive effect of the training to improve students' ability to monitor their own knowledge. Aleven and colleagues [1] describe Help Tutor, an agent integrated into a tutoring system that trains the help-seeking skill, an individual skill to seeking help from teachers, classmates, books, and others, when this help is necessary. This behavior is related to knowledge monitoring, since adequate help-seeking behavior suggests the student is able to identify what they know and what they do not know [23]. Other works [3, 18, 28] also encourage students to reflect on their knowledge at some point of the instruction, but their main goal is to train other metacognitive skills.

The work in [10] was the only one to investigate the effects of training to improve the knowledge monitoring skill, although it also trained other metacognitive skills in parallel. None of the studies we found investigated the knowledge monitoring skill exclusively, so the isolated effects of an instruction that specifically encourages students to reflect on their knowledge to improve the knowledge monitoring skill in computer learning systems are so far unknown. Additionally, an important characteristic of metacognitive training is adapted instruction [2]. None of the related studies we found adapts the portion of the instruction that encourages student's reflection on personal knowledge.

The goal of this study is to verify the effect of explicitly instructing the student to reflect on their knowledge in a computer learning systems through a training which encourages learners to reflect on their knowledge and has its content and frequency of intervention adapted to the characteristics of students. We aim at answering the following research questions: (RQ1) Does adaptive training improve the student's knowledge monitoring? (RQ2) Does adaptive training improve the student's performance in the domain?

The first hypothesis of the authors of this study is that the adaptive training provided effectively improves knowledge monitoring (H1), because the training encourages students to reflect on their knowledge, making them act less hastily, and also because it makes students aware of the importance of this metacognitive skill in their studies. Additionally, both frequency of intervention and intervention content are adapted to the current metacognitive skill of the student, their performance in the domain, and their problem solving history. The benefits of this adaptive characteristic of the metacognitive training have already been verified in classroom [2]. In the context of computer learning environments, existing research has used messages to lead the student to some reflection on their own knowledge. However, these works have not investigated the isolated effect of the students' reflection on their own knowledge. They also did not adapt instruction to the to the student's current level of knowledge monitoring. **The second hypothesis of the present work is that students' performance in the domain is improved by the training provided by the agent (H2)**, since the actions of the agent lead the student to be less hasty. Hence, the study also intends to verify whether a positive correlation may be found between performance in the domain and knowledge monitoring skill. This correlation has already been found in conventional learning environments [23, 24], but not in computer learning systems.

3 Knowledge Monitoring Training Agent

This Section presents the animated pedagogical agent proposed that trains the metacognitive skill of knowledge monitoring. The animated pedagogical agent encourages students to reflect on their knowledge. The following instructional strategies were adopted: (1) encouraging the student to identify what the problem is; (2) encouraging the student to dedicate some time to reflect on their knowledge before going to the next step in the problem solving process; (3) encouraging the student to reflect on similar, previously solved problems. These reflection strategies and their importance during problem solving are discussed in [9, 10, 16, 27].

The adaptive training is an important characteristic of the agent. Both training content and agent's frequency of intervention are adapted to the following information about the student: (1) current level of knowledge monitoring; (2) domain knowledge; and (3) problem solving history. In this adaptive context, the agent may be integrated into learning systems (for instance, ITSs) that: (1) provide step-by-step assistance during problem solving (e.g., step-based ITSs); (2) keep a problem solving record; (3) are able to identify the knowledge applied by learners in each step and the knowledge that might be applied in next steps; (4) infer students' knowledge probability in the domain knowledge units. The way the agent uses these characteristics of the learning system, which are fundamental to the training adaptation, is explained in following sections of this article.

The pedagogical agent uses three types of reflection actions, which are delivered either in the form of speech balloons, or simple text messages. Prompts encourage students to reflect on their knowledge, and occur before learners make an attempt at solving a new step. Feedbacks notify students of both inadequate behavior and their current knowledge monitoring level. Self-explanations encourage students to write in their own words how they monitored their knowledge.

3.1 Evaluating Knowledge Monitoring

To provide adaptive training, the pedagogical agent needs a mechanism to infer the students' current level of knowledge monitoring. This study used an instrument called Knowledge Monitoring Assessment (KMA), described in [23, 24], which measures learners' ability of monitoring their own knowledge. The KMA compares the student's assessment of their own knowledge to solve a problem with their actual performance on the same problem, generating a total of four possible scores: (a) the student assesses he has the knowledge required to solve the next step of the problem and effectively solves the problem (++); (b) the student assesses he has the knowledge to solve the next step of the problem, but cannot solve the problem correctly (+−); (c) the student assesses he does not have the knowledge required to solve the next step of the problem but effectively solves the problem (−+); and (d) the student assesses he does not have the knowledge required to solve the next step of the problem and does not solve the problem correctly (−−). Scores ++ and −− describe an accurate assessment by the student regarding his knowledge, whereas in +− and −+ cases, the assessment by the student was inaccurate. The scores consider only recent KMA assessments, so old metacognitive performance is not taken into consideration and only the current metacognitive state is assessed. The study only considers the last 15 KMA assessments, so more recent assessments contribute more than less recent assessments. Specifically, the most recent assessment contributes with 15 times its value, whereas the second most recent assessment contributes with 14 times its value. The fifteenth most recent assessment contributes with one time its value. Consider the following hypothetical scenario depicted in Table 1. In eight of the steps, the student assessed they would solve and did solve the problem, so score ++ for this student is 63 (15 + 12 + 11 + 8 + 6 + 5 + 4 + 2); in two of the steps the student assessed he would solve and did not solve the problem, so score +− for this student is 17 (14 + 3); in one of the steps the student assessed he would not solve and did solve the problem, totaling a value of 9 for score −+; finally, in four of the steps the student assessed he would not solve and indeed did not solve the problem, resulting in a −− score of 31 (13 + 10 + 7 + 1).

Table 1. Hypothetical scenario to assess KMA and its weights

	KMA assessments (from more recent ones to older ones)														
KMA	++	+−	−−	++	++	−−	−+	++	−−	++	++	++	+−	++	−−
Weight	15	14	13	12	11	10	9	8	7	6	5	4	3	2	1

The agent uses these scores to generate an index, called KMA index, which measures the discrepancy between students' self-assessment and their actual performance in the domain. This index reflects the learners' current level of knowledge monitoring. More specifically, this work uses the Hamman Coefficient (CH) formula (Formula 1), to calculate the KMA index, as suggested in [23]. This formula calculates the difference of the proportion of scores related to correct assessments (a and d) and of scores related to incorrect assessments (b and c).

$$CH = (((a + d) - (b + c))/((a + d) + (b + c))) \tag{1}$$

The KMA index is a real number between −1 and +1, in which +1 indicates precision in knowledge monitoring, while −1 indicates imprecision. The present study also classified the KMA index into two categories: satisfactory and unsatisfactory, separated by a threshold value of +0.5. Values that are equal or greater than the threshold express a satisfactory KMA. The choice of the threshold was made based on classroom pilot tests with the agent, and it aimed at making satisfactory KMA values high enough to express good and consistent metacognitive performance.

3.2 Operating Mechanisms

The agent has two operating mechanisms: (1) the inner flow and (2) the outer loop.

Outer Loop. Responsible for activating the inner flow. The outer loop is always executed before the student makes an attempt at solving a new step of the problem. The flow in Fig. 1 depicts the decision making process of the outer loop. Two strategies are used to decide whether the inner flow should be activated or not. The first strategy makes decisions based on student's metacognitive level, and it is used when the current KMA index of the learner is unsatisfactory, or when they have started the training too recently. A real number between −1 and +1 (i.e., the interval of the KMA index), is randomly generated using a function of uniform probability. If the generated value is equal or greater than the current KMA index of the student, the inner flow is activated. Therefore, the greater the KMA index, the smaller the probability of activation of the inner flow, i.e. the frequency of activation of the inner flow is adapted to the metacognitive level of the student.

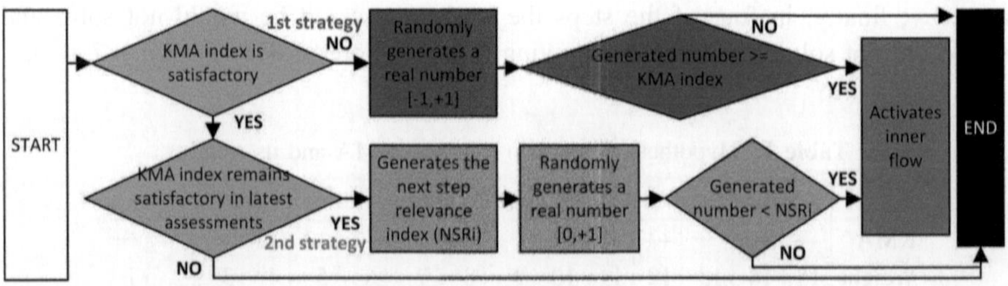

Fig. 1. Decision making process of the outer loop

The second strategy of the outer loop makes decisions based on the knowledge of the student in the domain, and it is used while the student keeps a satisfactory KMA index. Evidence found in pilot tests leads us to believe that students need to reflect more on their knowledge when they neither completely know nor completely do not know the solution to a problem. This evidence came from visual observations in combination with reports from the students during pilot tests. The observations and reports showed that many students spent longer thinking about their knowledge before taking a new step in problem solving when the next step requires knowledge with which they are familiar, but do not master. Accordingly, the second strategy of the outer loop uses the student's knowledge probability (real number between 0 and 1 inferred by the ITS) in each knowledge unit that can be used in the next step of the problem. A Knowledge Unit Relevance index (KURi) is calculated for each knowledge unit that the student can use in the next step of the problem. This study proposes an equation (Formula 2) to computes the KURi. Initially, the distance between the student's knowledge probability in the knowledge unit and the number 0.5 (50%) is calculated. Then, the ratio between this distance and the number 0.5 (50%) is calculated, and the complement of this ratio corresponds to the KURi.

$$1 - ((|0.5 - \text{knowledgeProb}|)/(0.5)) \tag{2}$$

The study also proposes the calculation of a Next Step Relevance index (NSRi), which is a real number between 0 and 1 that corresponds to the average between the KURis. A student with an NSRi close to 0 probably masters the topic or does not have the knowledge required for the next step. Relevance indexes close to +1 describe students who have an intermediate knowledge probability (close to 50%) of solving the next step. Finally, a random number (uniform probability) between 0 and +1 is generated. If this number is smaller than the NSRi the inner flow is activated, so it is more likely that the inner flow will be activated in steps in which the student needs to reflect more.

Inner Flow. Responsible for knowledge monitoring training. The flow in Fig. 2 depicts the decision making process of the inner flow. Initially, if the KMA index is unsatisfactory, the mechanism selects a prompt message that encourages students to reflect whether they have the required knowledge to solve the current step of the problem and waits for the solution. If learners are very hasty, that is, if they enter the step very quickly, an immediate feedback message is delivered, indicating inadequate behavior. As the flow indicates, before entering a new step, the students must assess whether they have the required knowledge to solve the step choosing "YES" or "NO". Next, the mechanism compares learners' self-assessment with their performance on the current step and updates the students' KMA index. Additionally, the mechanism may randomly choose to deliver a self-explanation activity in which students fill out a form describing the reasons for the assessment. Before each new step, the agent can also provide (randomly) feedback informing the learners' current level of knowledge monitoring.

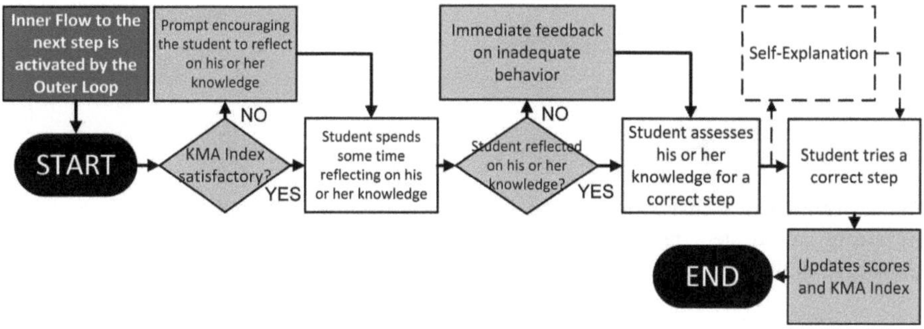

Fig. 2. Decision making process of the inner flow

3.3 Metacognitive Prompts

Prompts are text messages that are activated by the inner flow (as described in Sect. 3.2) and encourage students to reflect on their knowledge. The prompts were grouped into four levels. The first level encourages learners to reflect on their knowledge using the description of the problem. The second level makes students reflect on knowledge they have already demonstrated to master. The third level makes them reflect on solutions they provided previously that required knowledge that can also be applied to the current step. The fourth level shows a similar, previously solved step. Prompt levels are selected according to the students' KMA index. The smaller the current KMA index, the greater the prompt level that is selected (i.e., closer to level 4). In the evaluation, 65 messages were classified into four prompt levels to avoid that the system become predictable.

3.4 Implementing the Agent and Integrating It into the ITS

The APA is represented by a female, two-dimensional visual character called Pat. The APA is on the screen during the whole time that the student uses the system and has facial expressions related to speech and waiting mode, in addition to other expressions such as hand waves and winks. The agent was integrated into PAT2Math (Fig. 3a and b), a step-based ITS (available at http://pat2math.unisinos.br) that provides step-by-step assistance for students in the process of solving linear equations.

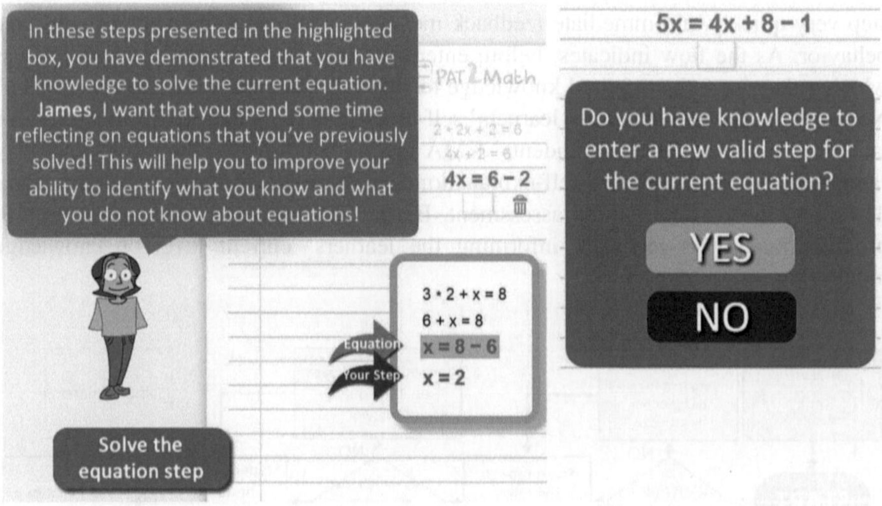

Fig. 3. Translated from Portuguese, (a) the agent delivering a prompt that shows a similar, previously solved step (prompt level 4), and (b) the agent asking the user to assess their knowledge in the new step

The content of the messages, including metacognitive prompt messages, metacognitive feedback and messages explaining the importance of using the metacognitive skill during the learning process, is delivered in speech balloons by the APA. In the self-explanation activity, which is one of the possible activities that the inner flow can deliver, students fill out a form in which they write, in their own words, their reasons to a given assessment of their own knowledge. The only goal of this activity was to encourage students to reflect on their metacognitive assessment.

4 Method

This Section provides the research design, participants, materials and procedures used in this evaluation study.

4.1 Design

This study followed an experimental design with control group and pretest/posttest. Different participants are used in each condition of the independent variable (no/yes metacognitive instruction). In the experimental group, the agent was configured for deliver metacognitive instruction, while in the control group, the agent was modified to not deliver metacognitive instruction, only providing hints related to the domain, which were also delivered by the agent of the experimental group.

The dependent variables were KMA index (as described in Sect. 3.1) and grades related to students' performance in pre and post-tests.

4.2 Participants

The experimental evaluation was carried out with three classes of eighth grade students and one class of seventh grade students in private schools in the south of Brazil. One hundred and seven students (ranging from 12 to 14 years old) participated in the study, but only the data of 63 students who provided the consent forms and participated in both pretest and posttest sessions were considered (12 students were from class 1, 18 students were from class 2, 22 students were from class 3, and 11 students were from class 4). In each class, students were randomly assigned to two groups: experimental and control. The control group was composed of 34 students (44% of the participants in this group were boys and 56% of the participants were girls), and the experimental group was composed of 29 students (55% of the participants in this group were boys and 45% of the participants were girls).

4.3 Materials

This study used the KMA instrument, as described in [23, 24], which measured learners' ability of monitoring their own knowledge. For each one of the two applications of the KMA instrument, in the pretest and posttest sessions, the authors of this work elaborated a sheet of paper containing ten items of first degree equations. For each equation student should answer whether he or she has knowledge to solve it or not. The authors also elaborated another sheet of paper containing the same ten items of equations for the student to solve. This instrument was used to collect student's metacognitive index (KMA index, as described in Sect. 3.1) and performance grades (integer numbers ranging from 0 to 10) in each application (as described in the next section) of the KMA instrument.

The animated pedagogical agent (as described in the Sect. 3) was integrated (as described in Sect. 3.4) into STI PAT2Math. In the experimental sessions, students used computers running one of the two different versions of the tutoring system with the agent.

4.4 Procedures

The experiment was composed of a total of six to seven sessions. In the first session, students went to the computer laboratory and received training in the tutoring system without the agent. In the second session, students returned informed consent forms (received in the previous session) signed by their parents and answered the pretest. In the following sessions (four sessions with two of the classes, and only three sessions in the other two classes, due to lack of schedule availability from the classes), students went to the computer laboratory of the school and solved equations in two different versions of the tutoring system with the agent. In the last session, students answered a posttest. Each session lasted for 50 min and all sessions had an interval of one week between each other.

The pretest and posttest sessions collected students' metacognitive indexes (KMA index) and performance grades in the domain. In the pretest session, students received a sheet of paper containing ten algebraic equations. For each equation, the student had to assess whether they had the knowledge to solve it or not. After returning the sheet of paper to the teacher, students received another sheet of paper containing the same ten equations, but this time they had to solve them. The same procedure was applied to the posttest session, but with different equations that had the same level of difficulty as the ones from the pretest. Students' assessments and performance were used to calculate the KMA index (as described in Sect. 3.1) for every student both in the pretest and in the posttest, together with the grades related to the performance of students in the domain.

Average and standard deviations were calculated in all experimental groups. One-way ANCOVA tests were conducted to compare KMA indexes and performance grades between groups. Student's t-tests were conducted to compare KMA indexes (pretest and posttest) within groups. The reliability of the data was attested using Ryan-joiner's test for normality and Levene's test for homogeneity of variances. Pearson's coefficient was conducted to verify the correlation between the KMA index and the performance in the domain. All the tests (alpha = .05) were executed in IBM SPSS Statistics and Statdisk.

5 Results

This Section presents averages and standard deviations from the control and experimental groups for all the data analyzed. It also presents the results of the statistical tests related to the KMA indexes (Sect. 5.1) and grades for performance in the domain (Sect. 5.2). Results from the correlation tests are also shown here (Sect. 5.3).

5.1 Knowledge Monitoring Skill

A paired samples t-test compared KMA indexes between the pretest and the posttest in each group, and did not find any statistically significant differences ($t(33 = .766$, $p = .225$) between the mean of the KMA index in the posttest ($M = .605$, $SD = .413$) and the mean of the same index in the pretest ($M = .544$, $SD = .325$) in the control group. However, a significantly higher mean ($t(28) = 3.333$, $p < .01$) was found in the posttest ($M = .800$, $SD = .251$) when compared to the pretest ($M = .569$, $SD = .381$) in the experimental group. A one-way ANCOVA test comparing the mean of KMA index in the experimental group ($M = .800$, $SD = .251$) with the mean of the control group (M = .605, SD = .413) in the posttest, using the pretest's KMA index from the experimental (M = .569, SD = .381) and control (M = .544, SD = .325) groups as covariate found a significant difference between groups (F(1.61) = 4.841, p = .032, partial $\eta 2$ = .075), indicating that the KMA index found in the experimental group is higher than the one found in the control group. Table 2 also presents these results.

Table 2. Knowledge monitoring skill and performance grade in the domain

Dependent variables	Control		Experimental		ANCOVA between groups	Paired samples t-test within groups	
	Pretest *M* *(SD)*	Posttest *M* *(SD)*	Pretest *M* *(SD)*	Posttest *M* *(SD)*	(posttest) *F(1.61)* *p* *partial η2*	Control *t(33)* *(p)*	Exper. *t(28)* *(p)*
KMA index	.544 (.325)	.605 (.413)	.569 (.381)	.800 (.251)	4.841 .032 .075	.766 (.225)	3.333 (< .01)
Performance grade	5.500 (2.034)	7.441 (2.205)	6.207 (2.242)	8.621 (1.741)	3.539 .065 .056	6.203 (< .001)	6.055 (< .001)

5.2 Performance in the Domain

A paired samples t-test compared the grades between the pretest (M = 6.207, SD = 2.242) and posttest (M = 8.621, SD = 1.741) in the experimental group and grades between the pretest (M = 5.500, SD = 2.034) and posttest (M = 7.441, SD = 2.205) in the control group. A significantly higher grade mean was found in the posttest both in the experimental condition (t(28) = 6.055, p < .001) and in the control condition (t (33) = 6.203, p < .001). A one-way ANCOVA test compared the grades of the students in both conditions in the posttest, using the performance grade in the pretest of the experimental (M = 6.207, SD = 2.242) and control (M = 5.500, SD = 2.034) groups as covariate. A difference with marginal statistical relevance (F(1.61) = 3.539, p = .065, partial $\eta 2$ = .056) was found between performance grades in the experimental (M = 8.621, SD = 1.741) and control (M = 7.441, SD = 2.205) groups, indicating that the performance in the domain in the group that received the metacognitive instruction was probably higher than the performance of students who did not receive the instruction. These results also are presented in Table 2.

5.3 Knowledge Monitoring Skill X Performance in the Domain

A Pearson's correlation coefficient analysis found a statistically significant positive correlation between the KMA index and the performance in the domain in the experimental condition. In the first test, which referred to the data collected during the pretest, control (r = +.45, p = .006) and experimental (r = +.60, p < .001) conditions presented a moderate positive correlation. In the second test, which referred to the data collected during the posttest, a strong positive correlation was found in the experimental condition (r = +.92, p < .001) and a weak positive correlation was found in the control condition (r = +.29, p = .090). These results are also presented in Table 3.

Table 3. Correlation between the KMA index and the performance grade

Correlation tested	Control *r (p)*	Experimental *r (p)*
KMA index (pretest) X Performance grade (pretest)	*+.45 (.006)*	*+.60 (< .001)*
KMA index (posttest) X Performance grade (posttest)	*+.29 (.090)*	*+.92 (< .001)*

6 Discussion

This Section analyzes the results presented in the previous Section. The goal is to discuss whether they provide evidence that supports the hypotheses from the study: the metacognitive agent's adaptive training improves the skills students have of monitoring their knowledge (H1); and the agent's adaptive training improves performance in the domain (H2). We also want to discuss whether it is possible to find a positive correlation between knowledge monitoring performance and performance in the domain.

The results of the statistical tests (see Sect. 5.1) presented positive evidence that supports hypothesis H1. All tests indicated that students in the experimental condition (i.e., who used the algebraic ITS that had the metacognitive training of the agent) had higher metacognitive indexes than students in the control condition (i.e., who did not receive the training of the agent). That is, results in the experimental group were significantly higher in the tests that compared the KMA index between groups in posttest and within groups (i.e., between pretest and posttest).

The results of the statistical test (see Sect. 5.2) showed that both groups had a significant improvement in performance in the posttest comparing with pretest. Because of the positive impact of our tutor on students' learning, shown in previous experiments [19], we already expected this result. Besides, the results also presented positive evidence that supports the hypothesis H2, i.e., the metacognitive training leads to higher learning gains. Higher posttest performance values were found in the experimental group than in the control group, with marginal statistical significance.

This study also found evidence (see Sect. 5.3) of a strong positive correlation between knowledge monitoring skill and learning by students that received training from the agent. All correlation tests between KMA indexes and performance grades were statistically significant, except for the second test in the control group. A strong positive correlation was found in the experimental group in posttest (second correlation test). However, the same tests presented a weak positive correlation in the control group. This positive evidence supports results found in classroom, as described in [23, 24]. These results had not been verified in computer learning systems, yet.

The correlation analysis does not infer causality, that is, it does not allow us to conclude that one variable causes the other, but it indicates the level of relationship when these variables are observed in the conditions of the evaluation. However, taking into consideration that the statistical tests indicate that students in the experimental group performed higher, it is possible that these results occurred because of the improvement in the metacognitive skill of knowledge monitoring, which was also higher in the experimental group. Thereby, these results on correlation analysis also support hypothesis H2 that the metacognitive instruction of the agent may improve performance.

We believe the system evaluation had a few limitations that impacted the results of the study. One of those limitations is that students only had three or four sessions of ITS use due to scheduling restrictions from the schools. We believe that results would have been even better if students had an extended period of ITS use, given that one of the fundamental principles to achieve success in metacognitive instruction is that the training of the metacognitive skills must be prolonged [26]. A longer training would allow evaluations on the effects of training for longer periods.

Due to restrictions in the calendar of the schools, it was not possible to make use of retention tests. If retention tests had been applied a few weeks after the posttest sessions, for instance, we would have been able to verify whether the knowledge monitoring skill persists after the agent's training.

Finally, this study also has limitations in regard to the generalization of its results. Although participants were randomly divided into control and experimental groups, which allows us to verify a causal relation between the cognitive training and the improvement in both knowledge monitoring skill and performance in the domain, it was not possible to use random sampling, which would allow us to generalize the results to other students. The sample used in the study is a convenience sample, involving students from three private schools in São Leopoldo, Brazil. Hence, the replication of the current study with other profiles of participants is important to generalize the results.

7 Conclusions

This work investigated the effects of adaptive training of the metacognitive skill of knowledge monitoring in an ITS, a computer learning systems that have intelligent modules that infer students' information. An animated pedagogical agent that trains knowledge monitoring in step-based ITSs was implemented. The agent encourages students to reflect on their knowledge during problem solving and adapts both training content and frequency of intervention to the students' current level of knowledge monitoring, their knowledge in the domain and their problem solving history in the ITS.

In related work, systems encourage students to reflect on their knowledge, but they train several metacognitive skills together. In that way, the present study intends to discover the specific effects of instruction that encourages students to reflect on their knowledge to promote knowledge monitoring in computer learning systems, as ITSs. Another important consideration is that in none of the existent works the knowledge monitoring training is adapted to the characteristics of the student.

In the evaluation, the animated pedagogical agent was integrated into PAT2Math, a step-based ITS that helps students when solving first degree equations. The results suggest that the adaptive training of the knowledge monitoring skill, which considers the current metacognitive skill of the student, his performance in the domain and his problem solving history in ITSs, may improve students' knowledge monitoring and learning. A positive correlation between knowledge monitoring and performance in the domain was also found when students received the agent's training.

When measuring the metacognitive skill of knowledge monitoring, the KMA does not take into an account the students' overconfidence level, in which the confidence a person has on their own judgment exceeds the accuracy of the judgment. This is a known cognitive bias in the assessment of knowledge monitoring. The current study also does not consider the affective states of the student, such as emotions, humor and engagement. An interesting future work is to investigate how overconfidence and affective states interfere in the knowledge monitoring skill and in the training of this skill, allowing the creation of new mechanisms to measure the metacognitive skill and the improvement of the training of this skill in computer learning systems.

The study also did not perform any treatment or evaluation with regards to the textual content generated by the student in the self-explanation activities delivered by the metacognitive instruction of the agent. Future work could analyze this content, relating it to the other actions of the student in the system and to the current level of the student in the knowledge monitoring skill, and also make inferences about engagement, humor and other affective states of the student. The results obtained by these analyses could be used to improve the metacognitive training and inference of the level of the knowledge monitoring skill.

The results of this study could help educational settings in basic education, such as elementary school and high school, to train the students' metacognitive ability of knowledge monitoring. The acquisition of this skill could positively affect other essential skills to students enter higher education, such as self-regulated learning skills. Students are less likely to engage in self-regulated learning activities such as planning, deploying study strategies and learning assessment, when they fail to identify what they know and what they do not know. Students acquire a great deal of knowledge during their basic education experiences that will be helpful on higher education contents, such as Calculus. Successful undergraduate students can self-monitor their understanding to recall their prior knowledge, to adjust their learning to concentrate on materials about unfamiliar contents and work harder in the development of their deficit areas.

Acknowledgments. The present work has received financial support of the following Brazilian research funding agencies: CNPq and FAPERGS.

References

1. Aleven, V., Mclaren, B., Koedinger, K., Rool, I.: Toward meta-cognitive tutoring: a model of help-seeking with a cognitive tutor. Int. J. Artif. Intell. Educ. **16**(2), 101–128 (2006)
2. Azevedo, R., Hadwin, A.F.: Scaffolding self-regulated learning and metacognition - implications for the design of computer-based scaffolds. Instr. Sci. **33**(5–6), 367–379 (2005)
3. Azevedo, R., Feyzi-Behnagh, R.F., Duffy, M., Harley, J.M., Trevors, G.: Meta-cognition and self-regulated learning in student-centered leaning environments. In: Jonassen, D., Land, S. (eds.) Theoretical foundations of Student-Center Learning Environments, pp. 171–197. Routledge, New York (2012)

4. Conati, C., Vanlehn, K.: Toward computer-based support of meta-cognitive skills: a computational framework to coach self-explanation. Int. J. Artif. Intell. Educ. **11**, 389–415 (2000)
5. Desoete, A., Roeyers, H., De Clercq, A.: Can offline metacognition enhance mathematical problem solving? J. Educ. Psychol. **95**(1), 188–200 (2003)
6. Dunlosky, J., Metcalfe, J.: Metacognition. SAGE, Los Angeles (2009)
7. Everson, H.T., Tobias, S.: The ability to estimate knowledge and performance in college: a metacognitive analysis. Instr. Sci. **26**(1–2), 65–79 (1998)
8. Flavell, J.: Metacognition and cognitive monitoring: a new area of cognitive-developmental inquiry. Am. Psychol. **34**(10), 906–911 (1979)
9. Fogarty, R.: The Mindful School: How to Teach for Metacognitive Reflection. IRI/Skylight, Palatine (1994)
10. Gama, C.A.: Integrating metacognition instruction in interactive learning environments. Ph. D. thesis, University of Sussex, Brighton (2004)
11. Goodman, B., Soller, A., Linton, F., Gaimari, R.: Encouraging student reflection and articulation using a learning companion. Int. J. Artif. Intell. Educ. 151–158 (1997)
12. Isaacson, R.M., Fujita, F.: Metacognitive knowledge monitoring and self-regulated learning: academic success and reflections on learning. J. Scholarsh. Teach. Learn. **6**(1), 39–55 (2006)
13. Kautzmann, T.R., Carlotto, T., Jaques, P.A.: Adaptive training of the metacognitive skill of knowledge monitoring in intelligent tutoring systems. In: Micarelli, A., Stamper, J., Panourgia, K. (eds.) ITS 2016. LNCS, vol. 9684, pp. 301–306. Springer, Cham (2016). https://doi.org/10.1007/978-3-319-39583-8_33
14. Kramarski, B., Mevarech, Z.R., Arami, M.: The effects of metacognitive instruction on solving mathematical authentic tasks. Educ. Stud. Math. **49**(2), 225–250 (2002)
15. Kramarski, B., Michalsky, T.: Student and teacher perspectives on IMPROVE self-regulation prompts in web-based learning. In: Azevedo, R., Aleven, V. (eds.) International Handbook of Metacognition and Learning Technologies. SIHE, vol. 28, pp. 35–51. Springer, New York (2013). https://doi.org/10.1007/978-1-4419-5546-3_3
16. Polya, G.: How to Solve It. Princeton University Press, Garden City (1957)
17. Puntambekar, S., Boulay, B.D.: Design and development of MIST - a system to help students develop metacognition. J. Educ. Comput. Res. **16**(1), 1–35 (1997)
18. Ramandalahy, T., Vidal, P., Broisin, J.: An intelligent tutoring system supporting metacognition and sharing learners' experiences. In: Aleven, V., Kay, J., Mostow, J. (eds.) ITS 2010. LNCS, vol. 6095, pp. 402–404. Springer, Heidelberg (2010). https://doi.org/10.1007/978-3-642-13437-1_84
19. Schaab, B., Duarte, M., Abs, D., Jaques, P.: Aplicação do mindfulness em um sistema tutor inteligente: um estudo piloto. In: Brazilian Portuguese, XXVI SBIE, vol. 26, no. 1, p. 1072 (2015)
20. Schoenfeld, A.H.: What's all the fuss about metacognition? In: Cognitive Science and Mathematics Education, pp. 189–215 (1987)
21. Stavrianopoulous, K.: Adolescent's metacognitive knowledge monitoring and academic help seeking. Coll. Stud. J. **41**(2), 444–453 (2007)
22. Tarricone, P.: The Taxonomy of Metacognition. Psychology Press, New York (2011)
23. Tobias, S., Everson, H.: Assessing metacognitive knowledge monitoring. In: Issues in the Measurement of Metacognition, pp. 147–222. Buros Institute of Mental Measurements and Erlbaum Associates, Lincoln, NE (2000)
24. Tobias, S., Everson, H.: The importance of knowing what you know: a knowledge monitoring framework for studying metacognition in education. In: Handbook of Metacognition in Education, pp. 107–127. Routledge, New York (2009)

25. VanLehn, K., et al.: The affective meta-tutoring project: lessons learned. In: Trausan-Matu, S., Boyer, K.E., Crosby, M., Panourgia, K. (eds.) ITS 2014. LNCS, vol. 8474, pp. 84–93. Springer, Cham (2014). https://doi.org/10.1007/978-3-319-07221-0_11
26. Veenman, M.J., Van Hout-Wolters, B.H.M., Afflerbach, P.: Metacognition and learning: conceptual and methodological considerations. Metacogn. Learn. **1**(1), 3–14 (2006)
27. Verschaffel, L.: Realistic mathematical modelling and problem solving in the upper elementary school: analysis and improvement. In: Teaching and Learning Thinking Skills, pp. 215–240 (1999)
28. Woolf, B.P., et al.: A general platform for inquiry learning. In: Cerri, Stefano A., Gouardères, G., Paraguaçu, F. (eds.) ITS 2002. LNCS, vol. 2363, pp. 681–697. Springer, Heidelberg (2002). https://doi.org/10.1007/3-540-47987-2_69
29. Zimmerman, B.J.: Investigating self-regulated and motivation: historical background, methodological development, and future prospects. Am. Educ. Res. J. **45**(1), 166–183 (2008)

Technology-Enhanced Solutions: Gamification and Educational Games

The Dark Side of Gamification: An Overview of Negative Effects of Gamification in Education

Armando M. Toda(✉), Pedro H. D. Valle, and Seiji Isotani

Institute of Mathematics and Computer Science (ICMC), University of São Paulo, São Carlos, São Paulo, Brazil
armando.toda@gmail.com, pedrohdvalle@gmail.com, sisotani@icmc.usp.br

Abstract. Gamification has a great number of studies in the education area since the emergence of the term. However, there is a lack of primary and secondary studies that explore the negative effects that gamification may have on learners, and lack of studies that analyze the gamification design that are linked to those negative effects. Based on this premise, we aim at answering the following research question "What are the negative effects that can occur in gamification when applied to educational contexts?". We seek to answer this question by analyzing the negative effects that are associated to gamification and the gamified learning design that are linked with them. To answer this question, we conducted a systematic mapping study to identify these negative effects. Based on the studies that were analyzed, we identified and mapped 4 negative effects and their gameful design. Loss of Performance was the most occurring effect and Leaderboard the most cited game design element, among other 11 elements. Moreover, elements and effects were linked in order to identify how these elements may have influenced on these outcomes. Based on our results, we found out that the game design may lead to a negative impact. For instance, Leaderboards are strongly associated to many negative effects mapped in this work. This result is corroborated by the psychology literature regarding ranking systems within learning environments. We believe our work may be useful to guide gamification instructors and specialists to avoid those negative effects in education contexts, by avoiding some game design elements settings.

Keywords: Gamification · Negative effects · Education

1 Introduction

Gamification has been widely explored in the past 7 years since the term was conceptualized by Deterding [13], as the use of game elements outside their scope. Since then, gamification studies have covered many areas, ranging from Economy to Education. However, most of gamification studies focus on the educational context [11], where researchers aim those game elements to improve learning

A. I. Cristea et al. (Eds.): HEFA 2017, CCIS 832, pp. 143–156, 2018.
https://doi.org/10.1007/978-3-319-97934-2_9

efficiency and learner's motivation [18]. Most of gamification studies in education have focused on exploring the positive motivational achievements however, there is still a gap regarding the identification of negative outcomes on students [6,14].

It is also well known that gamification is context-dependent, and inserting game elements such as Points, Badges, and Leaderboards (PBL Approach), without proper design, will not ensure the positive desired outcomes [5]. Gamification experts are already concerned about individual and collective variables, such as profiling users, in order to recommend a well-thought gamified scenario. Therefore, there is still a lack of studies that address or are concerned with the potential negative effects caused by gamification on learners in the educational context [19]. Thiebes et al. [33] present a list of issues that should be addressed by specialists when designing a gamified task or system, which are: Declining Effects, Cheating the System, Privacy, and Task Quality. However, those issues are related to online gamified environments, and not specifically to educational contexts. Kim and Werbach [19] debate some ethical issues related to gamification, such as Exploitation and Manipulation, presenting a framework on how to avoid them, again these issues are not related to education or learners. Finally, Andrade et al. [1] discuss some negative effects of gamification, such as lack of attention, in the Intelligent Tutoring Systems area. Other gamification studies also imply a set of properties that need to be addressed in the design of gamification in order to deploy it, especially regarding the Education area, such as learners' genres and player profiles [4,28].

In this context, this study aims at answering the following question: "What are the negative effects that can occur in gamification when applied to educational contexts?". To answer our research question, we performed a systematic mapping study, based on Kitchenham [20] guidelines, aiming at classifying the negative outcomes and identifying the gamification design. In this study, we consider a gamification design as a set of strategies using game elements to perform a task and a negative effect as a consequence that has a negative impact on the student. To present our research, this paper is organized as follows: Sect. 2 presents the related works, Sect. 3 presents the protocol. Section 4 presents the results and, finally, Sect. 5 presents the conclusions, discussions and future work.

2 Related Work

Gamification secondary studies have increased drastically in the past 7 years. Some of them, as the work performed by de Sousa Borges et al. and Dicheva et al. [6,14], focus on associating the good impact of gamification elements (considering their properties and rules) in educational contexts, in which most of the analyzed studies have achieved positive outcomes, dealing with users' behavior, and increasing their motivation. These authors also present a gap regarding the lack of empirical studies to attest the positive effects of gamification, and also do not address possible negative outcomes related to gamification bad design in education domain.

Another relevant study is conducted by Thiebes et al. [33], which focus at identifying which game elements were used in the field of Information Systems.

The authors propose a list of issues that can be influenced by gamification bad design however, the authors state that these are exclusive to online environments with gamified tasks.

Markopoulos et al. [22] also conducted a literature overview of gamification in education domain, presenting a list of critic points which correspond to extrinsic rewards that, when badly designed in gamified applications, can decrease learners' intrinsic motivation. The authors also claim that gamification may not be well accepted by teachers and instructors and that it can be difficult to obtain a balanced gamified scenario. Finally, they claim that gamification may not be applicable to every learning context.

In other domains, the positive effects of gamification are also not validated. According to Pedreira et al. [27] in their systematic mapping on gamification applied to Software Engineering, there is still a lack of empirical studies to attest the effects promoted by gamification. Furthermore, an analysis of those effects and how to design their gamified strategies is needed. This statement is reinforced by Sigala [30], where the author presents a literature overview regarding gamification in Marketing, Tourism, and Crowdsourcing. According to the author, gamification may "wear out" in a scenario in which gamification design has not been considered.

These studies allow us to observe a gap, in the sense that (i) negative outcomes of gamification are known but not properly mapped and (ii) there is a lack of information regarding negative effects and the gamification design that is used. Based on this gap, the next sections present the mapping of the negative effects associated with gamification in educational contexts, along with their gamification designs.

3 Protocol

The focus of this systematic mapping is to answer our main question "What are the negative effects that can occur in gamification when applied to educational contexts?" by identifying the main negative effects caused by gamification on learners. In order to conduct this study, we used the online tool Parsif.al and Google Docs to organize the mapping, as well as the guidelines proposed by Kitchenham [20], which consist of 3 steps: (i) planning, where we define our research questions, Inclusion and Exclusion Criteria and other definitions related to the mapping; (ii) conduction, where we conduct our research in the selected databases, select the papers and perform the data extraction; and (iii) results, where we summarize the data and evaluate the results. Based on these guidelines, we defined 2 research questions to guide this study:

RQ1 - What are the negative outcomes related to gamification in Education?

This question focus on identifying what are the main issues related to gamification applied to Education. By identifying those issues, we can encourage future studies on how to avoid them. Furthermore, the literature lacks formal classification of those issues regarding the educational context.

RQ2 - How is the gamification design related to these outcomes?

This question aims at associating the negative effects with the respective gamification designs, in order to identify which game elements and rules are linked to those effects. By identifying the gamification setup, we aim at encouraging future studies to carefully choose which and how those game elements may impact students' perception on gamification in the context of Education.

In order to answer these research questions, we defined 3 main terms to create the research string: Gamification, Education and Negative Effects. Some negative effects were previously collected based on 6 control papers that we used to aid in the string creation, which are: *gamification AND (failure OR problem OR privacy OR declining effect OR cheating the system OR exploit OR bad OR addiction) AND (education OR learning OR training OR instruction)*. This string returned all of the 6 control papers in one or more research bases.

As for the research bases, we choose the ones that are used and cited in Computer Science domain and were used in other secondary studies, as ACM Digital Library, AISEL, IEEE Xplore, ISI Web of Science, ScienceDirect, Scopus and Wiley Online Library. As for the Selection Criteria, we defined 4 Inclusion and 6 Exclusion Criteria. In our Inclusion Criteria (IC), we defined that (a) studies should have been published in the past 7 years; (b) studies should be in the English language; (c) studies should contain explicit evidence(s) regarding a negative effect associated with gamification; and (d) studies should be full papers published in a conference, journal or book chapter. As for our Exclusion Criteria (EC), we excluded (a) studies focused on serious games, since game development differs drastically from gamification design; (b) studies that were short papers, position papers or panels; (c) studies that contained risks associated with different areas from Education; (d) studies that did not contain any explicit negative effect associated with gamification; and (e) secondary studies.

Through the adequacy of the string to each of the databases, we began the searches in June, 2016. Our search returned a total of 1328 papers, that were screened by title, abstract and keywords. After this step, we obtained a set of candidate studies (c = 220 papers) that would be filtered by our Inclusion and Exclusion Criteria. After the application of the criteria and the deletion of duplicate papers, we obtained the final set of studies (n = 17 papers) that would be analyzed and discussed in this work. We also included 3 studies that were not returned by our search string but were cited in other papers. The number of studies and the complete list of analyzed papers can be seen on <https://goo.gl/3qrKv7>.

After the final selection, we began the Data Extraction phase, in which we mapped the negative effects into the gamification design that were used in the studies. These negative effects were based on their occurrence in the papers, by checking the exact words or the general meaning when analyzing the results of the primary studies.

4 Results and Discussions

This section presents the answers to our research questions based on our protocol. In Fig. 1, we analyze the areas of the selected studies. From this figure, we observe that most of the studies are in the field of Computer Science—specifically 10 studies. However, we also found studies in Administration, Communication, Mathematics, High School, Engineering and Arts.

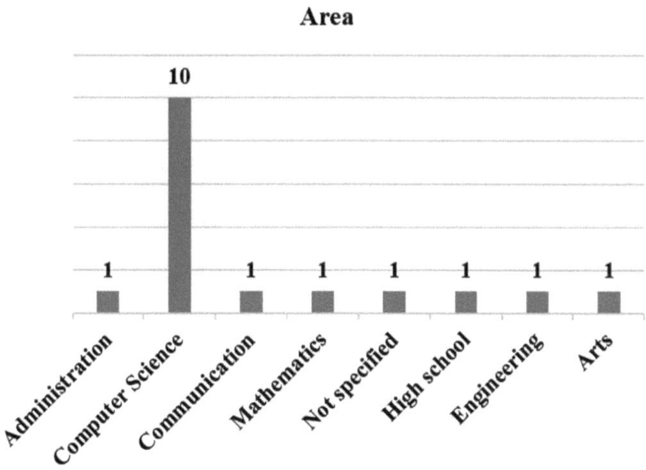

Fig. 1. Studies per area

We also analyzed the type of publication of the studies. From Fig. 2, we can observe that 10 studies were published in Symposiums/Conferences, 6 studies were published in Journals and only 1 study was published in a Workshop.

The selected studies are considered recent since all of them were published in the last 5 years. This behavior was already expected since gamification is considered a recent area of research. In Fig. 3, we can observe the distribution of the studies per year of publication. Most of the analyzed studies were published between 2014 and 2015.

We also analyzed the country of origin regarding the first author of each study, to verify which countries and research groups have been conducting research on the topic of this systematic mapping. We observed that most of the research has been carried out by researchers from the United States and Germany, as presented in Fig. 4.

Furthermore, by analyzing the focus of the studies, we identified 3 main ones: tool, where the authors described a technology to deploy the gamification; evidence, where the focus of the study was to give evidence regarding gamification properties; and method, where the authors focused on describing a method or process. The distribution of studies per focus is shown in Fig. 5. We observed that most studies focused on tools, specifically 9 of them.

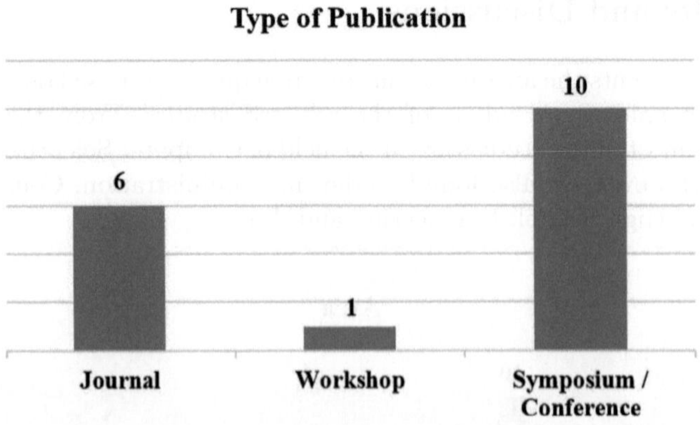

Fig. 2. Studies per type of publication

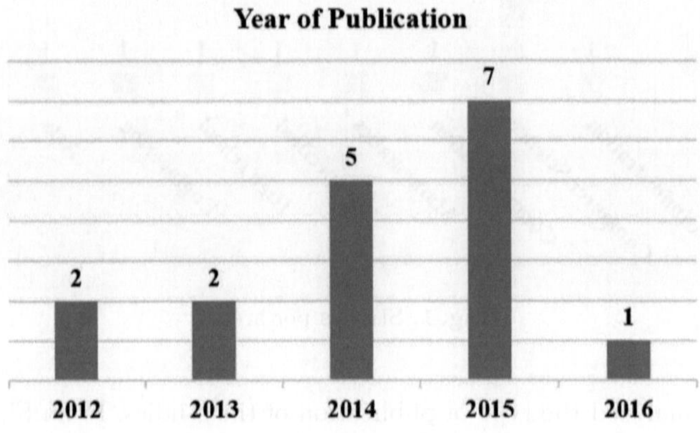

Fig. 3. Studies per year of publication

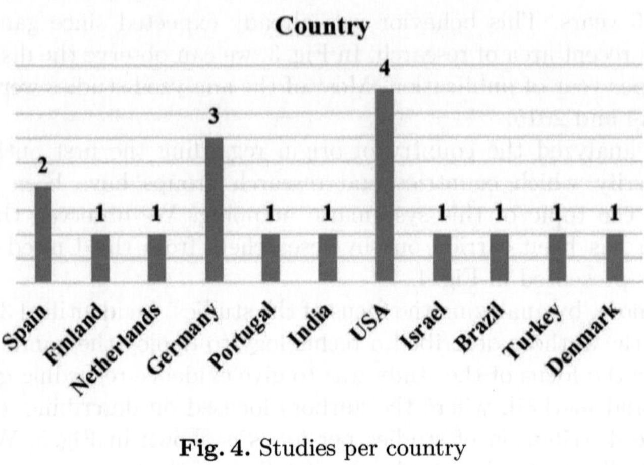

Fig. 4. Studies per country

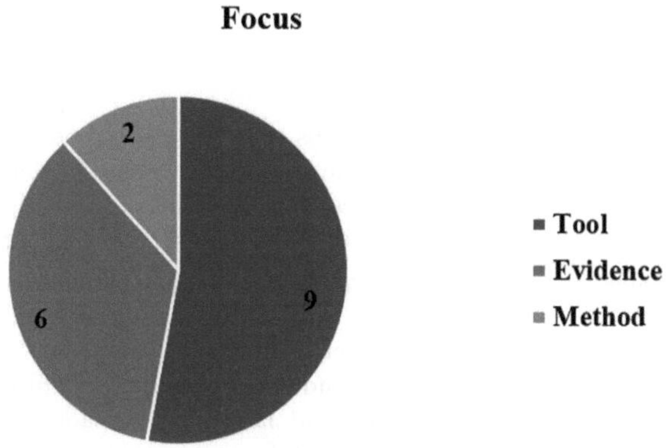

Fig. 5. Focus of the studies

Finally, regarding the outcomes of each analyzed study, we observed that 6 studies had positive results, 2 studies had negative results and 9 studies had mixed results, as presented in Fig. 6. This classification was based on the outcomes reported by the authors of the selected studies.

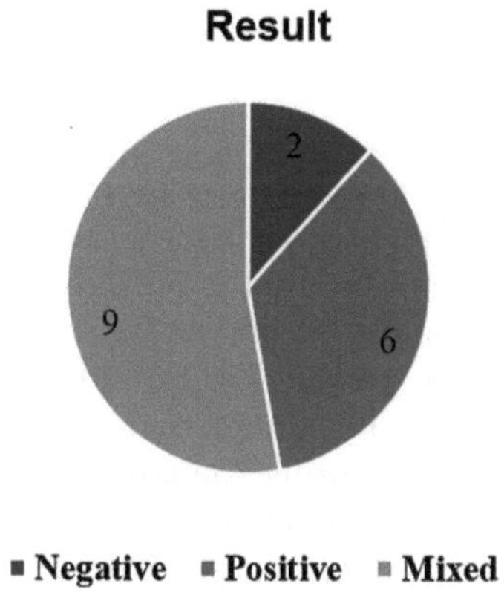

Fig. 6. Reported results of the studies

4.1 What are the Negative Outcomes Related to Gamification in Education?

Based on the outcomes and reports in the studies, we identified 4 negative effects: Indifference, Loss of performance, Undesired behavior and Declining effects. Those effects and their references can be seen in Table 1 below.

Loss of Performance was the most reported issue, identified in 12 studies. This issue arises from tasks and situations where gamification harms or hinders

Table 1. Negative effects and their studies

Effect	Frequency	References
Indifference	6	De-Marcos et al. [12], Haaranen et al. [16], Buisman and van Eekelen [7], Berkling and Thomas [4], Domínguez et al. [15], Papadopoulos, Lagkas and Demetriadis [26]
Loss of performance	12	De-Marcos et al. [12], Haaranen et al. [16], Barata et al. [3], Berkling and Thomas [4], Naik and Kamat [25], Hanus and Fox [17], Attali and Ariev-Attali [2], Campos et al. [8], Prause and Jarke [29], Kocadere and Çağlar [21], Snow et al. [32], McDaniels, Lindgren and Friskics [23]
Undesired behavior	9	Haaranen et al. [16], Singer and Schneider [31], Domínguez et al. [15], Codish and Ravid [10], Campos et al. [8], Prause and Jarke [29], Snow et al. [32], Papadopoulos, Lagkas and Demetriadis [26], McDaniels, Lindgren and Friskics [23]
Declining effects	5	Berkling and Thomas [4], Hanus and Fox [17], Domínguez et al. [15], Campos et al. [8], Attali and Ariev-Attali [2]

students' learning process. Most of those studies reported mixed outcomes, with positive and negative results after applying gamification in their learning contexts. De-Marcos et al. [12] reported that gamification may be related to loss of performance in the case of some students, and a similar scenario is seen in the study performed by Naik and Kamat [25]. Barata et al. [3], Attali and Ariev-Attali [2] and Hanus and Fox [17] stated that gamification had an influence on learners' loss of performance during the study due to demotivating effects. Campos et al. [8] reported that some learners did not understand the rules and this may have hindered their performance. There is a similar report made by Prause and Jarke [29], but in this study, learners did not like being penalized in the gamified activity. Snow et al. [32] reported that students that were more active in the gamified activity scored lower than their peers in the transfer skill tests. Kocadere and Çağlar [21] found similar results, as their learners were more focused on the gamified mechanics than on the assessment. Finally, McDaniels, Lindgren and Friskics [23] stated that students felt that the gamified activities were too difficult, which also impacted their grades.

Undesired Behavior was the second most cited issue, presented in 9 studies. This occurred because gamification caused a different effect (positive or negative) on the learning context in which it was applied, either due to bad planning or to the lack of it. Haaranen et al. [16] mentioned that the number of tasks declined in the gamified environment, which was the opposite of what the authors expected. Singer and Schneider [31] reported that some students stated that the bonus system were not satisfying, and they did not receive most of the gamified notifications due to problems in the system. The gamified application of Domínguez et al. [15] was not well received by learners, who stated that the system was not enjoyable and that it was a waste of time. Codish and Ravid [10] stated that their mechanics caused demotivation due to excessive competition, which was also reported in the studies of Campos et al. [8], Prause and Jarke [29] and Papadopoulos, Lagkas and Demetriadis [26], since they employed similar mechanics. Snow et al. [32] stated that learners were more focused on the gamification mechanics than on the learning assessment itself, when they should have done the opposite. Finally, McDaniels, Lindgren and Friskics [23] reported that the students became frustrated for not completing the badges, also stating that the activities were too difficult.

Following Undesired Behavior is Indifference. This occurs when gamification is reported to not have influenced, either for the better or for the worse, the learners within the study. De-Marcos et al. [12] reported that, overall, gamification did not improve learners' gain of knowledge compared to the traditional learning method. A similar thought is shared by the studies of Domínguez et al. [15] and Papadopoulos, Lagkas and Demetriadis [26], in which gamification did not exert any impact on cognition and performance. Haaranen et al. [16] stated that learners felt indifferent towards the gamified application, which for them was neither enjoyable nor boring. Buisman and van Eekelen [7] reported in their study that gamification did not impact learners' motivation and engagement. Finally, Berkling and Thomas [4] also reported that some students were not interested in the gamification that was implemented, choosing traditional methods over gamified ones.

Finally, Declining Effects are related to the gradual loss of motivation and engagement due to the gamification that was deployed, e.g. the students start the activity motivated due the novelty of gamification however it loses this motivation over time. This issue was found in 5 studies and, despite being similar to Loss of Performance from the viewpoint that learners' progress is hindered in both scenarios, it differs from Loss of Performance in the sense that learners' motivation and engagement decrease over time, which also may lead to Loss of Performance. Berkling and Thomas [4] stated that a group of students lost interest in the gamification over time as they used the system. Hanus and Fox [17] and Attali and Ariev-Attali [2] reported similar results, adding a gradual loss of motivation. Finally, Campos et al. [8] stated that students felt demotivated for not understanding the rules. Moreover, they became reluctant in the end of the study to use the gamified assessment, thinking that it could affect their grades.

Based on our findings from these studies, it is clear that those issues occurred due to the lack of proper methods and/or frameworks for planning and deploying gamification in a learning context. Another issue that influenced on those outcome is the absence of instructional theories to support the implantation of gamification, especially motivational design theories since most of the studies focused on increasing learners' motivation and engagement. These instructional design theories are needed to produce well-thought gamified strategies that will have positive impacts on the students. This statement is also corroborated by Attali and Arieli-Attali [2] and De-Marcos et al. [12].

4.2 How Is the Gamification Design Related to These Outcomes?

From the pool of selected papers, we were able to identify and classify 12 game elements that had been deployed: Leaderboard, Badge, Point, Level, Progression, Social Status, Social Interaction, Instant Feedback, Avatar, Economy, Challenge, and Narrative. We used the definitions regarding game elements based on the work of Dicheva et al. [14] and Thiebes et al. [33]. Other elements that were found were incorporated into the set of 12 elements due to similar semantics and different syntaxes. The distribution of elements per study can be seen in Table 2.

Table 2. Game elements and their frequency in the poll of studies

Game element	Frequency	Studies
Leaderboard	14	[3,4,7,8,10,12,15,17,21,23,25,26,29,31]
Badge	13	[3,4,8,10,12,15–17,21,23,25,31,32]
Point	12	[2–4,7,8,10,16,21,23,25,29,32]
Level	9	[2–4,10,12,21,25,29,32]
Progression	4	[4,10,12,25]
Social Status	4	[3,4,12,29]
Social Interaction	2	[29,31]
Instant Feedback	3	[4,21,32]
Avatar	3	[3,21,32]
Economy	1	[32]
Challenge	1	[4]
Narrative	1	[23]

Table 2 demonstrates a concentration on the PBL (Point-Badge-Leaderboard) approach, deployed by more than 10 studies. This may have impacted on most of the issues, since the PBL approach may not be suitable for certain situations and contexts [5]. This is especially the case when applying a gamification design template without considering individual profiles, instructional and motivational design theories. Next, we identified which elements were related to - and which ones impacted - the aforementioned negative effects. The results can be seen on Table 3. It is possible to observe that Leaderboard had a strong influence on almost all of the negative effects, followed by Point and Badge, both with the same influence.

Based on the mapping of the game elements into each negative effect in Table 3, we cannot affirm that these elements exert a high influence in causing

Table 3. Negative effects and their respective gamification designs

Negative Effect	# of elements	Elements	Most impacting element
Indifference	8	Leaderboard, Badge, Level, Progression, Social Status, Point, Instant Feedback, Challenge	Leaderboard and Badge
Loss of performance	11	Leaderboard, Badge, Level, Social Status, Social Interaction, Point, Avatar, Progression, Instant Feedback, Challenge, Economy	Leaderboard, Badge and Point
Undesired behavior	11	Leaderboard, Badge, Point, Level, Instant Feedback, Progression, Social Status, Social Interaction, Avatar, Economy, Narrative	Badge and Leaderboard
Declining effects	4	Leaderboard, Badge, Point, Level	Leaderboard and Point

these negative effects, but we believe that these findings demonstrate the necessity of studying those elements individually, in order to confirm that. However, this study allows us to state that deploying the PBL approach without proper instructional and motivational design support may lead to those negative effects, which is also debated by Bogost [5]. Proposals on this topic can be seen in the initial work of Toda et al. [24], in which the authors developed a metaprocess aiming at aligning lesson plans with gamification strategies in classroom scenarios and in the work of Chalco et al. [9], which proposes the use of ontologies to build proper gamified strategies in the context of Computer-Supported Collaborative Learning.

5 Conclusions, Limitations and Future Works

We proposed in this study a systematic mapping of the negative effects associated with gamification in educational contexts and aimed at understanding the relation among them. We believe that we were able to answer our research questions, as well as providing some advances to the existing literature by making an initial systematic classification of those negative effects in the context of Education. We also believe that this study may provide guidance to avoid undesired outcomes when applying gamification in educational scenarios, it corroborates the existing literature on the need of systematic methods and approaches that align instructional and motivational design theories with gamification-related objectives and strategies. Another important artifact from this study is the analysis between the negative effects and the main game elements that were used, which reinforces the idea that the PBL approach may cause various problems if not well designed or supported by instructional and motivational design theories.

This study, however, presents some limitations, one of them being the time frame in which the research was conducted, which excluded papers from the second semester of 2016. We also did not consider gray literature which might contain other significant results, nor papers from other languages rather than English.

In future work, we intend to carry out a deeper analysis of those primary studies, aiming at identifying their objectives and statistically analyzing how those negative effects influence each other. Individual analyses of each game element in the respective gamification designs are also in progress. Finally, we also intend to propose an approach to guide gamification experiments in order to avoid those negative outcomes, by means of a systematic process and ontological support.

Acknowledgments. The authors would like to thanks Capes, CNPq and FAPESP (project number 2016/02765-2) for the fundings of this project. We also would like to thanks the friends and researchers that aided in the review of the language of this paper.

References

1. Andrade, F.R.H., Mizoguchi, R., Isotani, S.: The bright and dark sides of gamification. In: Micarelli, A., Stamper, J., Panourgia, K. (eds.) ITS 2016. LNCS, vol. 9684, pp. 176–186. Springer, Cham (2016). https://doi.org/10.1007/978-3-319-39583-8_17
2. Attali, Y., Arieli-Attali, M.: Gamification in assessment: do points affect test performance? Comput. Educ. **83**, 57–63 (2015)
3. Barata, G., Gama, S., Jorge, J., Gonçalves, D.: Improving participation and learning with gamification. In: Proceedings of the First International Conference on Gameful Design, Research, and Applications, Gamification 2013, pp. 10–17. ACM Press, New York (2013)
4. Berkling, K., Thomas, C.: Gamification of a software engineering course and a detailed analysis of the factors that lead to it's failure. In: 2013 International Conference on Interactive Collaborative Learning (ICL), September 2013, pp. 525–530 (2013)
5. Bogost, I.: Gamification is bullshit. In: The Gameful World. MIT Press (2014)
6. de Sousa Borges, S., Durelli, V., Reis, H., Isotani, S.: A systematic mapping on gamification applied to education. In: Proceedings of the 29th Annual ACM Symposium on Applied Computing, SAC 2014, pp. 216–222 (2014)
7. Buisman, A.L.D., van Eekelen, M.C.J.D.: Gamification in educational software development. In: Proceedings of the Computer Science Education Research Conference, CSERC 2014, vol. 55, pp. 9–20 (2014)
8. Campos, A., Batista, E., Signoretti, A., Gardiman, R., Madeira, C.: Gamifying activities in a higher education course. In: Proceedings of the European Conference on Games-Based Learning, January–October 2015, pp. 117–124 (2015)
9. Challco, G.C., Mizoguchi, R., Isotani, S.: An ontology framework to apply gamification in CSCL scenarios as persuasive technology. Rev. Bras. Informática na Educ. **20**(2), 67 (2016)
10. Codish, D., Ravid, G.: Personality based gamification: how differente personalities perceive gamification. In: Proceedings of the 22nd European Conference on Information Systems, p. 11 (2012)
11. Darejeh, A., Salim, S.S.: Gamification solutions to enhance software user engagement - a systematic review. Int. J. Hum. Comput. Interact. **32**(8), 613–642 (2016). https://doi.org/10.1080/10447318.2016.1183330

12. De-Marcos, L., Domínguez, A., Saenz-de-Navarrete, J., Pagés, C.: An empirical study comparing gamification and social networking on e-learning. Comput. Educ. **75**, 82–91 (2014)
13. Deterding, S., Sicart, M., Nacke, L., O'Hara, K., Dixon, D.: From game design elements to gamefulness: defining "gamification". In: Proceedings of the 2011 Annual Conference Extended Abstracts on Human Factors in Computing Systems, CHI EA 2011, p. 2425 (2011)
14. Dicheva, D., Dichev, C.: Gamification in education: where are we in 2015? In: E-Learn 2015, Kona, Hawaii, USA, July 2014, pp. 1445–1454 (2015)
15. Domínguez, A., Saenz-de-Navarrete, J., De-Marcos, L., Fernández-Sanz, L., Pagés, C., Martínez-Herráiz, J.: Gamifying learning experiences: practical implications and outcomes. Comput. Educ. **63**, 380–392 (2013)
16. Haaranen, L., Ihantola, P., Hakulinen, L., Korhonen, A.: How (not) to introduce badges to online exercises. In: Proceedings of the 45th ACM Technical Symposium on Computer Science Education, SIGCSE 2014, pp. 33–38 (2014)
17. Hanus, M.D., Fox, J.: Assessing the effects of gamification in the classroom: a longitudinal study on intrinsic motivation, social comparison, satisfaction, effort, and academic performance. Comput. Educ. **80**, 152–161 (2014)
18. Kapp, K.M.: The Gamification of Learning and Instruction: Game-Based Methods and Strategies for Training and Education. Wiley, Hoboken (2012)
19. Kim, T.W., Werbach, K.: More than just a game: ethical issues in gamification. Ethics Inf. Technol. **18**(2), 157–173 (2016)
20. Kitchenham, B.: Procedures for performing systematic reviews (2004)
21. Kocadere, S.A., Çağlar, Ş.: The design and implementation of a gamified assessment. J. e-Learn. Knowl. Soc. **11**(3), 85–99 (2015)
22. Markopoulos, A.P., Fragkou, A., Kasidiaris, P.D., Davim, J.P.: Gamification in engineering education and professional training. Int. J. Mech. Eng. Educ. **43**(2), 118–131 (2015)
23. McDaniel, R., Lindgren, R., Friskics, J.: Using badges for shaping interactions in online learning environments. In: IEEE International Professional Communication Conference, pp. 12–15 (2012)
24. Toda, A., Silva, Y.R., Cruz, W., Xavier, L., Isotani, S.: Um processo de Gamificação para o ensino superior: Experiências em um módulo de Bioquímica. In: Anais do Workshop de Informática na Escola, p. 495 (2016)
25. Naik, V., Kamat, V.: Adaptive and gamified learning environment (AGLE). In: Proceedings of the IEEE 7th International Conference on Technology for Education, T4E 2015, pp. 7–14. IEEE (2016)
26. Papadopoulos, P.M., Lagkas, T., Demetriadis, S.N.: How revealing rankings affects student attitude and performance in a peer review learning environment. In: Zvacek, S., Restivo, M.T., Uhomoibhi, J., Helfert, M. (eds.) CSEDU 2015. CCIS, vol. 583, pp. 225–240. Springer, Cham (2016). https://doi.org/10.1007/978-3-319-29585-5_13
27. Pedreira, O., García, F., Brisaboa, N., Piattini, M.: Gamification in software engineering - a systematic mapping. Inf. Softw. Technol. **57**(1), 157–168 (2015)
28. Pedro, L.Z., Lopes, A.M.Z., Prates, B., Vassileva, J., Isotani, S.: Does gamification work for boys and girls? In: Proceedings of the 30th Annual ACM Symposium on Applied Computing, SAC 2015, pp. 214–219 (2015)
29. Prause, C.R., Jarke, M.: Gamification for enforcing coding conventions. In: Proceedings of the 2015 10th Joint Meeting on Foundations of Software Engineering, ESEC/FSE 2015, pp. 649–660. ACM Press, New York (2015)

30. Sigala, M.: Gamification for crowdsourcing marketing practices: applications and benefits in tourism. In: Garrigos-Simon, F.J., Gil-Pechuán, I., Estelles-Miguel, S. (eds.) Advances in Crowdsourcing, pp. 129–145. Springer, Cham (2015). https://doi.org/10.1007/978-3-319-18341-1_11
31. Singer, L., Schneider, K.: It was a bit of a race: gamification of version control. In: 2012 2nd International Workshop on Games Software Engineering (GAS), pp. 5–8 (2012)
32. Snow, E.L., Allen, L.K., Jackson, G.T., McNamara, D.S.: Spendency: students' propensity to use system currency. Int. J. Artif. Intell. Educ. **25**(3), 407–427 (2015)
33. Thiebes, S., Lins, S., Basten, D.: Gamifying information systems a synthesis of gamification mechanics and dynamics. In: Twenty Second European Conference on Information Systems, pp. 1–17 (2014)

An Experience with Software Engineering Education Using a Software Process Improvement Game

Daniela C. C. Peixoto[1(✉)], Rodolfo F. Resende[2], and Clarindo Isaías P. S. Pádua[2]

[1] Federal Center for Technological Education of Minas Gerais, Belo Horizonte, Brazil
cascini@decom.cefetmg.br
[2] Federal University of Minas Gerais, Belo Horizonte, Brazil
{rodolfo,clarindo}@dcc.ufmg.br

Abstract. Educational games offer different ways for representing complex themes. However, creating a game that precisely addresses the subject being taught and effectively provides students' learning and engagement is a complicated task. With this in mind, we researched and created an educational game for improving students' Software Process Improvement (SPI) knowledge. The aim of this game is to bring practices from the software development industry to the students. In an attempt to provide a better guidance, this work describes the issues and challenges associated with the creation and validation process of a Software Engineering educational game. We adopted an incremental and iterative approach, where each step involves different knowledge, allowing us to point out a set of important aspects that should be taken into account during the development. These aspects can guide new developers and instructors in the design and selection of educational games.

1 Introduction

One of the biggest challenges faced by society and, in particular, by universities is the need to make the process of people training more effective and efficient. The expectation of higher quality in education and training has not been satisfied in general and there is an increasing dissatisfaction in the industry related to the students' preparation in various economic sectors [1]. In addition, it is difficult to practice the concepts presented in the classroom by means of traditional teaching methodology, centered on the professor [2]. One way to improve this scenario is applying methods and technologies to support teaching, for example, activities developed typically in industrial projects, residence programs, and educational games.

Considerable changes in educational processes are expected with the use of new pedagogic and educational technologies. Face to face courses, centered on professors, which are commonly offered by worldwide universities, are being substituted by more dynamic courses, using hybrid innovative technologies. Important examples include the various free of charge online courses (*MOOC - Massive*

A. I. Cristea et al. (Eds.): HEFA 2017, CCIS 832, pp. 157–173, 2018.
https://doi.org/10.1007/978-3-319-97934-2_10

Open Online Course). Other examples include the use of educational games as a complementary method for teaching some topics addressed in undergraduate courses [3–5].

The use of educational games[1] represents a shift from "learning by listening" to "learning by doing" model of teaching [6]. The educational games offer different ways for representing complex themes. With them, students can practice activities that are infeasible to practice during an academic course, due to restrictions of time and resources. In the Software Engineering field, there are a number of researchers that discuss the use of educational games as educational tools [7,8]. Educational games in a Software Engineering curriculum suitably fits as a complementary component to lectures, projects, and readings [3,9]. Using educational games, it is possible to practice some Software Engineering skills in a more realistic environment, increasing students' motivation and engagement [7,10,11]. It is also possible to reduce the risks of failure and the high costs of an actual practice. In addition, students can work at their own pace.

Besides the great advance in the research of games as learning tools, the theoretical base of educational games is still incipient [12]. Here we describe our experience in developing and evaluating a Software Engineering educational game, highlighting the activities that were carried out and the challenges that we faced during the creation process. The game is SPIAL, acronym for *Software Process Improvement Animated Learning Environment*, which is a desktop-based, graphical, interactive, and adaptable educational game. The presented results can be used as an aid in the design of new games and to analyze the development processes of existing educational games. The main contribution is the set of aspects in the development and evaluation of SPIAL, showing the problems that we faced and the proposed solutions.

The remainder of this work is structured as follows. Section 2 outlines a discussion about the main aspects presented in this research. Section 3 presents an overview of the steps carried out during the educational game design and development. Section 4 shows how SPIAL was evaluated. Section 5 presents a discussion of issues and challenges faced during SPIAL development. Section 6 concludes this work with the final considerations and future work.

2 Background

2.1 Educational Games

Over the last years, there has been an upsurge of educational games research (e.g. [7,10,11]). In spite of differences among studies focuses and approaches [11], the results suggest that students enjoyed and had fun when playing these games. Students also agreed that games can support learning [13]. Connolly and and colleagues [7] identified, with an extensive literature review, that the most frequent outcomes in serious games are knowledge acquisition and content understanding. In the Software Engineering field, we observed similar outcomes.

[1] In this work we refer to educational games also using the term "serious games".

Students' perceptions about these games were favorable. For them, playing a game was an enjoyable experience.

Table 1 is based on two studies [9,14]. The table identifies 13 Software Engineering educational games. According to these studies, games in Software Engineering can be divided in two groups of topics: (i) software processes topics, including games whose final goal is to develop a software project within a certain set of constraints, and their rules are based on the Software Engineering lessons, such as SESAM [4], SimSE [3], and MO-SEProcess [15]; and (ii) specific knowledge, including games whose final goal and rules are specific to each game context, such as, The Incredible Manager [16], SimVBSE [17], and qGame [18].

Despite the increasing interest in Software Engineering games, a few frameworks have been proposed in order to help designers during the creation process [19]. The Van Staalduinen and de Freitas [20] work defines a game-based learning framework which can be used to design new games or to evaluate existing ones. It is structured in the steps of learning, instruction and assessment. The Marcos and Zagalo [21] work introduces a development model of serious games based on the processes of digital arts. The model encompasses six phases: Concept Design, Narrative Design, Experience Design, Game Design, Game Implementation, and Game Deployment Planning. Sommeregger and Kellner [22] proposed six design stages for the creation of educational adventure games, including conceptual design, game design, implementation, testing, validation and project management. Kirkley and colleagues [23] developed a model which supports a specific instructional system design. This model defines a formative and a summative evaluation. Loh [24] describes a 10-step instructional development model. It is a detailed model which includes: determining target audience and learning content, determining the amount of funding and time available, writing game narratives, selecting the game development kits/game bundle, designing the video game and the game mechanics, designing the interactive learning instruction, developing, beta testing and usability testing, public release, and efficacy assessment. Kickmeier-Rust and colleagues [25] work defines an interdisciplinary approach in order to design educational games. This model focus is on the psychological and pedagogical elements, such as analysis of curricula and definition of the background story according to the educational aims.

Assessment of learning effectiveness in educational games employs a mixed-method approach, including pre and post domain specific tests, questionnaires, interviews, and video analysis. Usually these methods cover few characteristics of the game environment, varying according to the needs and experiences of its designer [9]. Calderón and Ruiz [10] identified through a systematic literature review that the main methods for evaluation are questionnaires, and the main research interests are to determine effectiveness in terms of knowledge acquisition. The work of Petri and von Wangenheim [26] identified through a systematic literature review seven approaches to evaluate educational games. They verified that the majority of these approaches collected data via questionnaires. Most of the researchers of educational games agree that it is needed more investment in research in order to improve the approaches for evaluating educational games.

Table 1. An overview of some computer-based educational games (Extracted from Peixoto and colleagues [9] and Souza and colleagues [14])

Game	Teaching subject	Player role
SimSE [3]	Software engineering concepts	Project manager
SESAM [4]	Software engineering concepts	Project manager
The incredible manager [16]	Project management	Project manager
MO-SEProcess [15]	Software engineering concepts	Team member
SimVBSE [17]	Value-based software engineering	Project manager
qGame [18]	Requirements engineering	Project manager
iThink [27]	Requirement elicitation	Requirement engineer
SimSys [28]	Software engineering concepts	Software engineering student
SDM [29]	People management	Project manager
Ameise [30]	Software project management	Project manager
Venture [31]	Social factors, cultural and linguistic differences in GSD	Software engineer
eRiskGame [32]	Risk management	Project manager
RCAG [33]	Requirements collection and analysis	Several roles

2.2 SPI Educational Game

SPIAL [5,34] (*Software Process Improvement Animated Learning Environment*) is a graphical, interactive, and adaptable educational game (Fig. 1). The game goal is to improve the Software Engineering learning, using simulation. Several Software Engineering topics are explored within the context of a Software Process Improvement (SPI) project. These aspects were based on CMMI-DEV version 1.3 [35]. CMMI was chosen because it is the most widely known SPI reference model.

SPIAL was developed with the goal to provide students with a more realistic experience in software development processes within the academic environment. SPIAL allows students to practice SPI techniques and the best practices of Software Engineering. SPIAL is a single-player game in which the player takes on the role of a manager of an SPI group in a software development organization. SPIAL's scope covers both a development project and an improvement project. The player is given a process improvement task and he or she can interact with

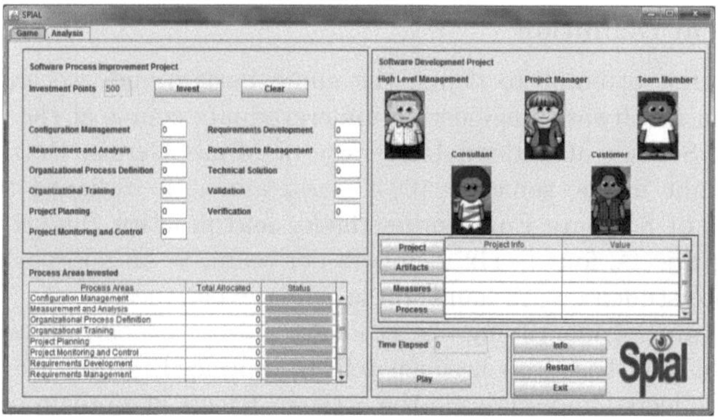

Fig. 1. SPIAL graphical user interface.

other stakeholders (high level management, project manager, team member, consultant, or customer) represented as non-player characters, i.e. a character controlled by the computer (see Fig. 1). The type of improvement that is required is stated at the beginning of the game and it can include, for example, cost and defect reduction and productivity improvement. In order to complete the task, player can make investments for improving specific process areas in a software development project. A good investment strategy will result in improvement of process areas and a bigger budget for further investments. The player can visualize project estimations, indications of process areas capability level and decide in which process area to invest. During the development project, the player can visualize the effects of his/her selections on the outcomes (productivity, defect, cost, and time-to-market measures) and, if needed, change his/her investments. The final outcome is a score that represents how close the results are to the initial proposed target. During the game, the non-player characters communicate the effects of the player's actions through bubbles over their heads.

3 SPIAL Creation

We used information about other Software Engineering educational games as a reference in SPIAL development. In fact, we developed SPIAL using the best practices we have found on the literature in order to avoid repeating old mistakes. For instance, we embraced different stakeholders' perspectives during the validation step.

No doubt that many researchers did not explore part of the activities presented by the educational games development processes, during the development stage. Indeed, as many other Software Engineering educational games, we did not follow a specific educational game development process. We incrementally and iteratively tried to cover the needed steps for SPIAL development during the requirement, design, development, testing and validation phases. Within each iteration, we executed a number of activities, including definition of the problem and the game environment, implementation and validation.

3.1 Problem Definition

In our very first attempt to define the game background, we analyzed data obtained from a Software Engineering undergraduate course at the Department of Computer Science at Federal University of Minas Gerais, Brazil [36]. This course is taught in one semester (60 h class) and it is designed to cover the fundamentals of Software Engineering theory and practice. The course adopts team-based projects, four to six members per team, as an approach to provide undergraduate students with hands-on learning experience.

With the aim of identifying the students' needs, we carried out a contextualized evaluation based on artifact defects produced during the team-based project. We collected and analyzed the artifact defects of 15 team projects from four classes, during two years. We also interviewed instructors and students in order to collect some qualitative impressions about the project main problems. After conducting these analyses, we observed that there was an undesirable gap between what was taught in the classroom and what the students had to do in the team project. As expected in an introductory course, it was not possible to provide the students all the details needed for an adequate software development, however it seemed that some important points were not made explicit.

Although this study was not replicated in other universities, we observed that, in general, there is a lack in the students' preparation, mainly regarding bridging the gap between Software Engineering course and the industrial software development environment.

With this scenario, we defined our main goal with SPIAL: overcome the problems identified above and contributed to the improvement of Software Engineering Education. Specifically, we expected that SPIAL would help the students to be able to:

- Practice concepts related to technical and managerial processes of software development.
- Reinforce the concepts learned in class, mainly, the Software Engineering lessons, carrying out a Software Process Improvement initiative. Specifically, the game rewards correct decisions (appropriate Software Engineering practices) and penalize unjustifiable ones.
- Follow the stages of a software development process in an organization.
- Learn some of the events that can occur in a project (e.g. resistance to change, lack of high-level management commitment).
- Observe possible results of improvement actions taken during development (e.g. the reduction of defects after performing the Requirement Engineering activities).
- Analyze the effects of following an immature software development process (e.g. more restricted process visibility, higher number of defects). Therefore, reinforcing the possible advantages of having a defined process.

3.2 Design

The primary concern of educational game design is to elaborate a system that is congruent with the learning goals. Designers must tell users what they mean by the educational game they have created, and players are expected to understand how to play the educational game and respond to it. Since designing is a crucial task in the game development, we evaluated the existing Software Engineering educational games in order to identify the main design decisions and incorporate the ones that match SPIAL's goals.

An assessment of the educational games [5], shows that although they are developed with different aims (e.g. SimSE x SimVBSE), they have a considerable number of common design aspects, for example, (i) clear goals and rules (ii) "trial and error" strategy, mainly, using feedback; (iii) clear definition of the virtual world where the game occurs; (iv) analysis of the adaptation of the learning environment according to different levels of initial knowledge, or different expectations and objectives; and (v) some game features.

We identified these as the core design elements and, below, we defined how they were incorporated into SPIAL.

- **Clear Goals and Rules:** A premise of SPIAL development was to have clear goals and rules. At the beginning of the game the player is given a process improvement task and a brief description of SPIAL learning goals. Therefore the players can understand their objectives and manage to achieve them. The rules were structured according to the Software Engineering and SPI best practices. The rules and their effects are presented gradually during the game play. In addition, the game provides a search functionality for the rules descriptions which can be triggered during the whole game.
- **Feedback:** After defining the context of our educational game as an SPI game, we envisaged how we could integrate the feedback. We have chosen to follow the same strategy of other educational games: Non-Player Characters (NPC), i.e. a character controlled by the computer, communicate the effects of the player's actions through bubbles over their heads. Therefore, the player can immediately visualize the effects of his/her actions.
- **Virtual World:** SPIAL also presents the metaphor of a physical software development organization, as in many other Software Engineering educational games. The player acts as an SPI manager, who can interact with other stakeholders (high level management, project manager, team member, consultant, or customer) represented as NPC.
- **Adaptability:** Since an essential feature of an educational game is to be easily adaptable, we designed a configurable simulation system in order to allow many definitions of simulation models. This characteristic impacts in the very first design decisions. This was an important characteristic to address distinct models of different processes, different process improvement frameworks, students with different learning styles and levels of initial knowledge, and instructors with different expectations and objectives.
- **Game Features:** We incorporated some common game features collected from the Software Engineering educational games, which include, life-like challenges (constraints on budget, defects and time), a score in a 10 points scale, random events, dialogue (textual messages), and interaction.

After design definitions, we started SPIAL development: we designed the user interface and the storyline, including the challenges and the game settings. We also defined the events and their effects on the game world.

3.3 Implementation

The knowledge of Software Engineering educational games and their common requirements, and the lack of support to educational games development led us to the decision of creating our own framework, FASENG (*Framework for Software Engineering Simulation Games*) [37]. FASENG is composed of three components[2] (simulation model, simulator, and simulation engine) that can be reused in new educational games creation (see Fig. 2). These components can also be found in other Software Engineering educational games [4,38], however, in our case, they can be reused in new developments[3].

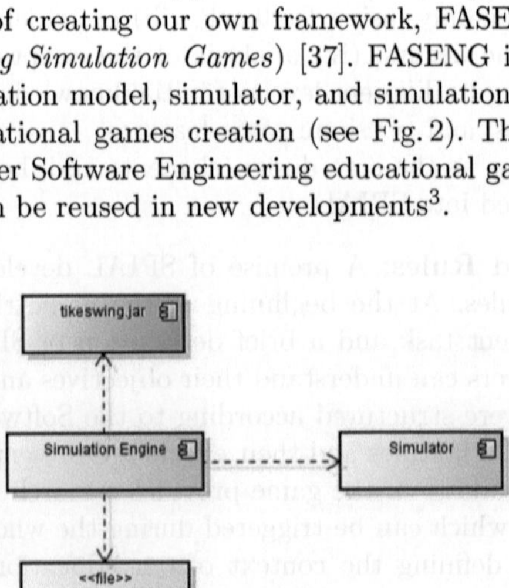

Fig. 2. FASENG structural elements.

The first configurable component is the **simulation model** that can be used to represent:

- The structure of the software process, activities, interactions and artifacts;
- The system behavior and the factors that influence the environment to be simulated;
- The initial scenario of the game, i.e. the initial state of the project to be simulated.

In the XML file, instructors can tailor a simulation model according to their course. This model allows the representation of an active role to players, reaction to events, feedback and different participants to actions.

The second component is the **simulator** which takes as input the model and interprets it, computing its equations iteratively. The simulator's final target is

[2] We use the term "components" to refer to the structural elements of FASENG.

[3] TikeSwing is a framework for Java Swing development providing high-level MVC architecture.

to calculate the behavior of each element. In the simulator component, aspects of discrete events are used to represent individual objects (e.g. process areas, and artifacts), actions ("give training or mentoring", "make an investment"), conditions of triggers and destroyers, and the rules executed at the start or at the end of an action. Characteristics involving continuous values are represented through rules for continuous domains, which are performed at each step of the simulation, whether the actions that define them are currently active. The simulator runs in a loop. For each time step, it looks whether the conditions of triggers and destroyers were satisfied, executes continuous rules, generates log, and gives feedback to the players.

The last component is the **simulation engine**, where the player interacts, receiving visual feedback from the simulation results and through which the player changes the model parameters. SPIAL has mainly three interfaces: the introductory interface (Fig. 3), the game tab sheet (Fig. 1) and the analysis tab sheet (Fig. 4). The introductory interface (Fig. 3) presents a description of the required knowledge to play the game, the game task, the player's role, the description of the software development company, the goals, how much time and money the player has, the final score calculation and some guidance to play the game. The game tab sheet (see Fig. 1) displays the software process improvement project, the stakeholders of the software development company and some project information. This is the main interface of the game, where important interaction occurs. The analysis tab sheet (see Fig. 4) was designed with the aim to giving players a better insight of the correctness of their decisions, specifically providing graphical information about the measures, the description of the rules used in the game, information about the achievement or not their improvement targets, and the final score.

4 Evaluation

The early versions of SPIAL were not evaluated in a controlled way, for instance, using formal experiments. Mainly the students were asked to report their impressions about the game in a subjective way. This approach was similar to those applied in most evaluations of educational games. In our last attempt, we organized the evaluated dimensions in a more structured way [9]. The goal was to allow the application of different techniques in order to receive feedback from different sources rather than only students' opinion.

Part of our approach was inspired by SimSE [8], since it is an example of educational game with extensive evaluation of the learning process it promotes. We evaluated SPIAL using three methodologies supported by UGALCO framework [9]: (1) Semiotic Inspection Method; (2) Experiment; and (3) User Test. These evaluations are detailed in Peixoto and colleagues [9] and briefly described below.

In the first evaluation, using the Semiotic Inspection Method (SIM) [39], an inspector carried out the assessment, playing the role of a student. After conducting the SIM steps, a unified analysis was produced, highlighting the main

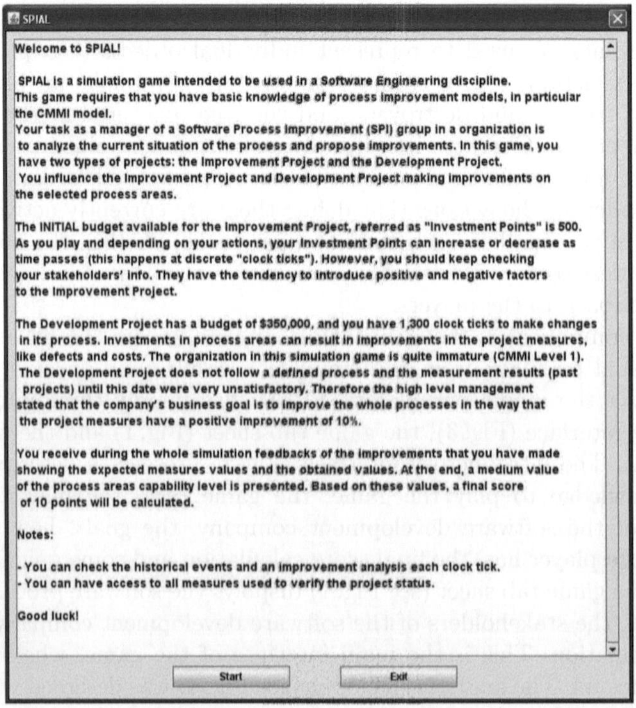

Fig. 3. SPIAL introductory screen.

communication breakdowns. According to the inspector, important aspects to understand the core behavior of the play were missing, such as the reason why sometimes investments do not produce any improvement. We produced a new game version, with some modifications, in order to provide more guidance to players.

Then, in the second evaluation, we carried out two experiments with undergraduate Computer Science students of a Software Engineering course. In the first experiment, our aim was to gain a better understanding of SPIAL effectiveness as an educational tool. We also determined the strengths and weaknesses of SPIAL through the feedback of the students who played it. In total, 11 students participated in this pilot experiment. Each student answered a background and a pre-test questionnaire, before playing SPIAL, and a post-test, after playing it. In general, the students were able to understand how the educational game works, they found SPIAL quite enjoyable, and they had fun during the game play. They learned new concepts and reinforced concepts taught in the Software Engineering course.

In the second experiment, we followed UGALCO framework [9] in order to evaluate the game experience, adaptivity, learning experience and usability dimensions. Students answered specific questions for each dimension. With respect to the Understandability/Usability dimension, an example of question is

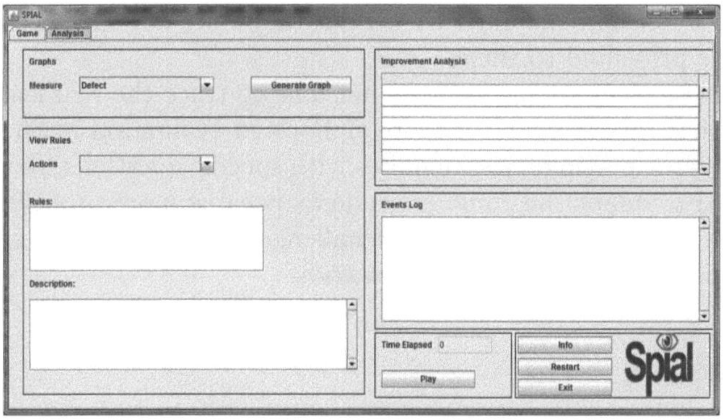

Fig. 4. SPIAL analysis tab sheet.

'how easy was the understanding of the game inputs and the outputs?'. The students' answers were mapped to a value range R = [0, 1], where a "strong negative" is encoded as "0", a "weak negative" as "0.25", undecided as "0.5", "positive" as "0.75", and a "strong positive" as "1". In total, 15 students participated in this experiment. The highest scored attribute was the learnability, i.e. the students understood how the game play works (Table 2). The attribute with smallest scored value was the learning goals, and this reflects the fact that players did not feel that they gained a great amount of new knowledge.

Table 2. Average results for UGALCO'S attributes.

Dimension	Attribute	Average
Game experience	Challenge	0.47
	Competence	0.68
	Immersion	0.70
	Positive experience	0.65
Adaptivity	Cognitive and motivational aspects	0.60
Learning experience	Learning goals	0.46
	Content appropriateness	0.63
Usability	Operability	0.45
	Understandability	0.63
	Learnability	0.77
	Attractiveness	0.47
	Satisfaction	0.62

The user test occurred in a one-to-one setting - one subject and one evaluator. As corroborating with the previous experiments, students reported that they reinforced SPI concepts playing SPIAL, but they were not sure about what new concepts were presented to them.

Besides these evaluations, we were not able to trace the real learning effects considering these assessments. A more elaborated evaluation mechanism, using pre and post game playing questionnaires with specific questions, should be done. We also failed to adapt the game as students became more proficient. Students reported that they were not anymore challenged and interested in playing the game after discovering its rules and behavior.

5 Discussion

The primary purpose of SPIAL is to match the learning objectives of an introductory Software Engineering course in an effective, efficient and engaging way. The design and development process was iterative and involved a number of different skills and activities, including the selection of the conceptual content, modeling of the real world processes and employment of suitable game-like elements.

Basically, for SPIAL creation we followed the main activities of traditional software development processes such as RUP (*Rational Unified Process*) [40]: requirement, design, development, and testing. After a long period of SPIAL development, implementation and validation, we considered successful the game creation process. The students' feedback provided evidence that the game can yield a positive effect to the learning of Software Engineering and SPI concepts. However, we identified that some activities could be carried out in a different way. Below, we revised the stages of SPIAL development describing what we could consider in a different perspective in a new game development.

5.1 Problem Definition

The conceptual content of SPIAL was designed around the SPI, Software Engineering rules, and Software Engineering process concepts. The concepts covered by the game were identified observing students' recurrent problems in a restrict academic environment. We believe that this analysis was sufficient for the development of a version that meets the needs of our course. However, a more comprehensive analysis should be conducted in order to provide support for the game adoption in other contexts.

In SPIAL, we identify the core of most widespread processes worldwide from the analysis of SPI initiatives reported by organizations in research papers. In most of the research studies, the context, where the improvement happens, was not defined. This brought some difficulties during the reuse of information and restricted the number of phenomena represented into the educational game. We observed this problem when we conducted our first experiment. Five students (45%) were undecided regarding the real aspects of an SPI initiative presented in SPIAL, five students (45%) felt that they learned only basic new concepts (i.e. not advanced concepts) of SPI, and six students (55%) reported that they only learned the basics about the real world of SPI initiatives in organizations.

Therefore, it was needed a more comprehensive evaluation of which real world phenomena should be covered by the educational game. This will make the simulation closer to the real software development organizations.

5.2 Design

As developers of instructional tools, designers need to address a set of characteristics and to deal with complex technologies. A framework for developing educational games could help in this difficult task. The educational games frameworks would help to clarify what kind of learning outcomes the students can expect, the identification of core game attributes, and the activities to be carried out. In this work, we showed that even without employing these frameworks, we achieved satisfactory results. However, following them:

- We could better define skill assessment mechanisms and embed them on the game. In SPIAL, we basically include the final score;
- We could first validate the storyline, before the implementation. It is important to execute activities that trace back concepts, processes and game-like elements during the design step;
- We could have better planned the integration of the game like features. Students reported that represented elements on the game interface should be improved and we realized that this influenced the students' learning and motivation.

5.3 Implementation

We created SPIAL using the FASENG framework. For its usage, it is required a basic Software Engineering knowledge and the definition of the game's requirements and design elements. We observed that more investigation is needed in this area, since at the time that we developed SPIAL, we did not find suitable support. To date, most of the Software Engineering educational games are developed from scratch. It is also important to create a tool to help modeling the game rules, for example, supporting FASENG designers in order to define the game rules.

5.4 Evaluation

During our research on educational game frameworks, we identified that most of them emphasize the initial development phases (conceptual project and design), and they did not invest sufficiently in different types of tests or assessments. This is also true in the development process of other Software Engineering educational games [28,32]. In these games, the assessments are mostly restricted to small pilot tests or subjective evaluations [9]. Therefore, we recognized the need of a more effective understanding of the educational contributions, mainly regarding the learning outcomes. In the literature, we found relatively few studies that have definitively assessed the effectiveness of the educational game. Well-rounded assessments can provide the basis for the improvement of this research field.

Students were confused regarding the real aspects of an SPI initiative presented in SPIAL. Although students learn more as they play more, the evaluation reveals that a longer playing time contributed to a feeling of repetitiveness. This happens mainly because the level of challenge did not change. This seemed to become a seriously detracting factor that was recurrent in our evaluations. At the end of this study, even with UGALCO, we have difficulties to track the real learning effect and to get statistically significant results. After SPIAL evaluations, we can affirm that there is a need to invest in the definition of more effective learning evaluation mechanisms.

6 Conclusion

This work describes the creation process of a Software Engineering educational game. Our experience with SPIAL has highlighted some important aspects that can be used to enhance the field of Software Engineering education and educational game development. The most concrete result is the educational game itself, along with its SPI simulation model. SPIAL was developed using a reusable framework, which allows its adjustment to distinct learning environments. This can assist students with different learning styles, different levels of initial knowledge, and instructors with different expectations and objectives. Other important results are the problems that we faced, which include: (i)there is a need of more investment in educational games assessments; (ii) improvement of the game development frameworks, suitably addressing the evaluation stages; (iii) analysis of the different needs, including instructors, students and other university stakeholders; and (iv) the demanding effort needed to calibrate the simulation. In the Software Engineering area, it is difficult to find enough data in the literature or with appropriate quality level.

As a future work, we plan to propose a game development process that will incorporate these relevant issues. We are working on improvements related to the UGALCO and FASENG frameworks. We are planning to conduct more evaluations and ask other instructors to participate in the experiments. These experiments will be supported with more specific guidelines. We are also evaluating different approaches for assessing students' skills and learning.

References

1. Aasheim, C.L., Li, L., Williams, S.: Knowledge and skills requirements for entry-level information technology workers: a comparison of industry and academia. J. Inf. Syst. Educ. **20**(3), 349–356 (2010)
2. Chen, W.F., Wu, W.H., Wang, T.L., Su, C.H.: A game-based learning system for software engineering education. In: 38th ASEE/IEEE Frontiers in Education Conference, Saratoga Springs, New York, October 2008

3. Navarro, E.O.: SimSE: a software engineering simulation environment for software process education. Ph.D. thesis, Donald Bren School of Information and Computer Sciences, University of California, Irvine (2006)
4. Drappa, A., Ludewig, J.: Simulation in software engineering training. In: 22nd International Conference on Software Engineering, ICSE 2000, Limerick, Ireland, 4–11 June 2000, pp. 199–208 (2000)
5. Peixoto, D.C., Resende, R.F., Pádua, C.I.P.: The issues of adopting simulation games in software engineering classes. In: 45th ASEE/IEEE Frontiers in Education Conference, El Paso, Texas, pp. 1–8, October 2015
6. Garris, R., Ahlers, R., Driskell, J.E.: Games, motivation, and learning: a research and practice model. Simul. Gaming **33**(4), 441–467 (2002)
7. Connolly, T.M., Boyle, E.A., MacArthur, E., Hainey, T., Boyle, J.M.: A systematic literature review of empirical evidence on computer games and serious games. Comput. Educ. **59**(2), 661–686 (2012)
8. Hainey, T., Connolly, T.M., Boyle, E.A., Wilson, A., Razak, A.: A systematic literature review of games-based learning empirical evidence in primary education. Comput. Educ. **102**, 202–223 (2016)
9. Peixoto, D.C.C., Resende, R.F., Pádua, C.I.P.S.: Evaluating software engineering simulation games: the UGALCO framework. In: 44th ASEE/IEEE Frontiers in Education Conference, Madrid, Spain, pp. 1–9, October 2014
10. Calderón, A., Ruiz, M.: A systematic literature review on serious games evaluation: an application to software project management. Comput. Educ. **87**, 396–422 (2015)
11. Boyle, E.A., et al.: An update to the systematic literature review of empirical evidence of the impacts and outcomes of computer games and serious games. Comput. Educ. **94**, 178–192 (2016)
12. Bedwell, W.L., Pavlas, D., Heyne, K., Lazzara, E.H., Salas, E.: Toward a taxonomy linking game attributes to learning. Simul. Gaming **43**(6), 729–760 (2012)
13. Caulfield, C., Xia, J.C., Veal, D., Maj, S.P.: A systematic survey of games used for software engineering education. Mod. Appl. Sci. **5**(6), 28–43 (2011)
14. de A. Souza, M.R., Veado, L., Moreira, R.T., Figueiredo, E., Costa, H.: A systematic mapping study on game-related methods for software engineering education. Inf. Softw. Technol. **95**, 201–218 (2017)
15. Ye, E., Liu, C., Polack-Wahl, J.: Enhancing software engineering education using teaching aids in 3-D online virtual worlds. In: 37th ASEE/IEEE Frontiers in Education Conference, FIE 2007, Milwaukee, Wisconsin, pp. T1E-8–T1E-13, October 2007
16. de Oliveira Barros, M., Dantas, A.R., Veronese, G.O., Werner, C.M.L.: Model-driven game development experience and model enhancements in software project management education. Softw. Process Improv. Pract. **11**(4), 411–421 (2006)
17. Jain, A., Boehm, B.: SimVBSE: developing a game for value-based software engineering. In: 19th Conference on Software Engineering Education and Training, Oahu, Hawaii, pp. 103–114, April 2006
18. Knauss, E., Schneider, K., Stapel, K.: A game for taking requirements engineering more seriously. In: First International Workshop on Multimedia Requirements Engineering, Barcelona, Catalunya, Spain, pp. 22–26. IEEE Computer Society (2008)
19. Ahmad, M., Rahim, L.A., Arshad, N.I.: A review of educational games design frameworks: an analysis from software engineering. In: 2014 International Conference on Computer and Information Sciences, Kuala Lumpur, Malaysia, pp. 1–6, June 2014

20. Van Staalduinen, J.P., de Freitas, S.: A game-based learning framework: linking game design and learning outcomes. In: Learning to Play: Exploring the Future of Education with Video Games, vol. 53, pp. 29–45 (2011)
21. Marcos, A., Zagalo, N.: Instantiating the creation process in digital art for serious games design. Entertain. Comput. **2**(2), 143–148 (2011)
22. Sommeregger, P., Kellner, G.: Brief guidelines for educational adventure games creation (EAGC). In: IEEE Fourth International Conference on Digital Game and Intelligent Toy Enhanced Learning (DIGITEL), pp. 120–122, March 2012
23. Kirkley, S.E., Tomblin, S., Kirkley, J.: Instructional design authoring support for the development of serious games and mixed reality training. In: Interservice/Industry Training, Simulation and Education Conference (2005)
24. Loh, C.S.: Researching and developing serious games as interactive learning instructions. Int. J. Gaming Comput.-Mediat. Simul. (IJGCMS) **1**(4), 1–19 (2009)
25. Kickmeier-Rust, M.D., Schwarz, D., Albert, D., Verpoorten, D., Castaigne, J.L., Bopp, M.: The Elektra project: towards a new learning experience. In: M3 - Interdisciplinary Aspects on Digital Media & Education, pp. 19–48 (2006)
26. Petri, G., von Wangenheim, C.G.: How to evaluate educational games: a systematic literature review. J. Univers. Comput. Sci. **22**(7), 992–1021 (2016)
27. Fernandes, J., Duarte, D., Ribeiro, C., Farinha, C., Pereira, J.M., da Silva, M.M.: iThink: a game-based approach towards improving collaboration and participation in requirement elicitation. Procedia Comput. Sci. **15**, 66–77 (2012)
28. Longstreet, C.S., Cooper, K.: A meta-model for developing simulation games in higher education and professional development training. In: 17th International Conference on Computer Games (CGAMES), pp. 39–44, July 2012
29. Kohwalter, T.C., Clua, E.W.G., Murta, L.G.P.: SDM - an educational game for software engineering. In: Brazilian Symposium on Games and Digital Entertainment (SBGAMES), pp. 222–231 (Nov 2011)
30. Bollin, A., Hochmuller, E., Mittermeir, R., Samuelis, L.: Experiences with integrating simulation into a software engineering curriculum. In: IEEE 25th Conference on Software Engineering Education and Training, pp. 62–71, April 2012
31. Monasor, M.J., Vizcaíno, A., Piattini, M.: Providing training in GSD by using a virtual environment. In: Dieste, O., Jedlitschka, A., Juristo, N. (eds.) PROFES 2012. LNCS, vol. 7343, pp. 203–217. Springer, Heidelberg (2012). https://doi.org/10.1007/978-3-642-31063-8_16
32. Galvão, T.A.B., Neto, F.M.M., Bonates, M.F., Campos, M.T.: A serious game for supporting training in risk management through project-based learning. In: Putnik, G.D., Cruz-Cunha, M.M. (eds.) ViNOrg 2011. CCIS, vol. 248, pp. 52–61. Springer, Heidelberg (2012). https://doi.org/10.1007/978-3-642-31800-9_6
33. Hainey, T., Connolly, T.M., Stansfield, M., Boyle, E.A.: Evaluation of a game to teach requirements collection and analysis in software engineering at tertiary education level. Comput. Educ. **56**(1), 21–35 (2011)
34. Kupsch, D.C.C.: SPIAL: a tool for software process improvement training. Ph.D. thesis, Universidade Federal de Minas Gerais, Brazil (2012)
35. CMMI: CMMI®for development, version 1.3. Technical report CMU/SEI-2010-TR-033, Software Engineering Institute (2010)
36. Peixoto, D.C.C., Batista, V.A., Resende, R.F., Pádua, C.I.P.S.: Learning from students' mistakes in software engineering courses. In: 40th ASEE/IEEE Frontiers in Education Conference, FIE 2010, Northern Virginia/Washington, D.C., pp. F1J-1–F1J-6, October 2010

37. Peixoto, D.C.C., Possa, R.M., Resende, R.F., Pádua, C.I.P.S.: FASENG: a framework for development of software engineering simulation games. In: 42nd ASEE/IEEE Frontiers in Education Conference, FIE 2012, Seattle, Washington, D.C., October 2012
38. Dantas, A.R., Barros, M.O., Werner, C.M.L.: A simulation-based game for project management experiential learning. In: 16th the International Conference on Software Engineering and Knowledge Engineering. SEKE 2004, Banff, Anbert, Canada, pp. 19–24, June 2004
39. de Souza, C.S., Leitão, C.F., Prates, R.O., Amélia Bim, S., da Silva, E.J.: Can inspection methods generate valid new knowledge in HCI? The case of semiotic inspection. Int. J. Hum Comput Stud. **68**, 22–40 (2010)
40. Kruchten, P.: The Rational Unified Process: An Introduction, 3rd edn. Addison-Wesley Professional, Boston (2004)

Using Ontology and Gamification to Improve Students' Participation and Motivation in CSCL

Geiser Chalco Challco[1(✉)], Riichiro Mizoguchi[2], and Seiji Isotani[1]

[1] ICMC, University of São Paulo, São Carlos, SP, Brazil
geiser@usp.br, sisotani@icmc.usp.br
[2] Japan Advanced Institute of Science and Technology, Nomi, Ishikawa, Japan
mizo@jaist.ac.jp

Abstract. During Collaborative Learning (CL), scripted collaboration may cause a motivation problem which makes the students to dislike and drop out the group activities over time. In order to cope with such a problem, we proposed the use of gamification as a technology to increase the students' motivation and engagement in CL scenarios. As consequence of increasing the motivation of students, we assume a reduction in the desistance of CL activities when the scripted collaboration is gamified. However, gamification is a complex task that requires knowledge about game elements (such as leaderboards and point systems), game design (e.g. how to combine game elements) and their impact on motivation and learning. To address these issues, we have developed an ontology that aims to give structured guidance on how to gamify CL scenarios. In the study presented here, we focused on describing the ontology and how it is used to gamify CL scenarios. We demonstrate the effectiveness of this approach to deal with the motivation problem through an empirical study in which the students enrolled in the introduction to computer Science course at the university of São Paulo participated in either non-gamified CL sessions or ontology-based gamified CL sessions. Significant differences were found in the students' intrinsic motivation and the percentage of students by groups who had incomplete participation in the CL session. The students who participated in ontology-based gamified sessions reported to be more intrinsic motivated than the students who participated in non-gamified CL sessions, and the percentage of students by groups who had incomplete participation was significantly less in ontology-based gamified sessions than in non-gamified CL sessions. These results indicate that our approach can be used to deal with the motivational problem caused by the scripted collaboration. They also validate the assumption that the gamification of CL scenarios can be used to reduce the desistance of students in CL activities when CSCL scripts are used as a method to orchestrate and structure collaboration.

Keywords: Gamification · Ontology · Motivation · Collaboration scripts CSCL

A. I. Cristea et al. (Eds.): HEFA 2017, CCIS 832, pp. 174–191, 2018.
https://doi.org/10.1007/978-3-319-97934-2_11

1 Introduction

One common problem in distance learning education is the high percentage of students who drop out their courses, modules and learning activities. In particular, during the collaborative learning (CL) activities, when the interaction among the participants is orchestrated and structured by Computer-Support Collaborative Learning (CSCL) scripts, the students are not motivated to participate in these scenarios over time [7]. The CSCL scripts are the technology that describes the way in which the participants of a CL scenario should interact to attain a set of pedagogical goals defined by an instructional design, and it has been demonstrated to be useful to support meaningful interaction in the CL process [13, 18, 26]. Thus, in the past few years, efforts of the CSCL community have been directed to finding solutions that increases both students' motivation and learning outcomes through the use of CSCL scripts [8, 19, 20].

In this direction, several researchers and practitioners have investigated Gamification as a promising technology to deal with the motivation problem in different learning contexts [28, 29]. Gamification as "*the use of game design elements in non-game contexts*" [6] aims to increase the students motivation and engagement through the introduction of game elements, such as points, leaderboards and so on. In this sense, the benefits that a student gets through the gamification of a learning scenario strongly depends on how well these game elements are applied and how well they are linked with the pedagogical approaches employed in the learning scenario [17]. Therefore, the educational benefits in a CL session, e.g. the reduction of students who drop out CL sessions, only will happen when there is an adequate design [28]. Nevertheless, such task is not trivial, many current uses of gamification are incorrect or poorly designed [31]. The main reason for such poor design includes the assumption that the same game design element can be shared for different situations. For example, a leaderboard with individual rankings is assumed that makes the scenario more enjoyable for everyone. However, this leaderboard is only enjoyable for a participant who likes the competition and for a participant who wants to be on the first positions.

To deal with these issues, we have developed an ontology that adequately represent the organization of the knowledge related to the CL design and gamification. By ontology, we refer to a representation of entities, properties and relations of a target-domain in a manner that can be understood by human and computer system agents. With this formalization, the ontology can be used by humans and intelligent theory-aware systems as guidelines to personalize the gamification of CL scenarios, and thus, to deal with the motivation problem caused by the scripted collaboration. The next section describes the ontology and how it is used to gamify CL scenarios. Section 3 presents the formulation of the empirical study conducted to validate our approach. Section 4 presents the obtained results in the empirical study, and the interpretation of these results. Finally, Sect. 5 presents the related works, final remarks, and future steps.

2 Ontology to Gamify CL Scenarios

The *Ontology to Gamify Collaborative Learning Scenarios* - called OntoGaCLeS (available at website: http://ontogacles.caed-lab.com) - uses the model of roles [23] to represent knowledge extracted from the theories and practices related to gamification, and how these concepts are applied in CL scenarios to deal with the motivation problem caused by the scripted collaboration. This formalization consists in a set of ontological structures that has been developed as an extension of *CL Ontology* [14].

In this section, after to present an overview of the CL ontology, we detail the ontological structures proposed to describe the gamification of CL scenario, and at the end of section, we show how these structures can be used by humans and intelligent theory-aware systems to personalize the gamification of CL sessions.

2.1 CL Ontology: Ontological Structure to Represent CL Scenarios

The *CL ontology* has a long history of development, with contributions from several researchers, and has been successfully applied to support the modeling of learners' development [12], the interaction analysis [11], and the design of CL sessions [13].

The CL scenarios and their elements are represented in the ontology using the concepts and terms shown in Fig. 1(a), where: **I-goal** is the individual learning goal of the student in focus "*I*." **I-role** and **You-role** are the CL roles played by the students "*I*" and "*You*;" **Y <= I-goal** is the learning strategy used by the student "*I*" to interact with the student "*You*" in order to achieve his/her individual learning goals (*I-goal*); **W(L)-goal** is the common learning goal for group members; and **W(A)-goal** is the rational arrangement of the group's activity used to achieve the common goal for group members (*W(L)-goal*) and individual learning goals (*I-goal*).

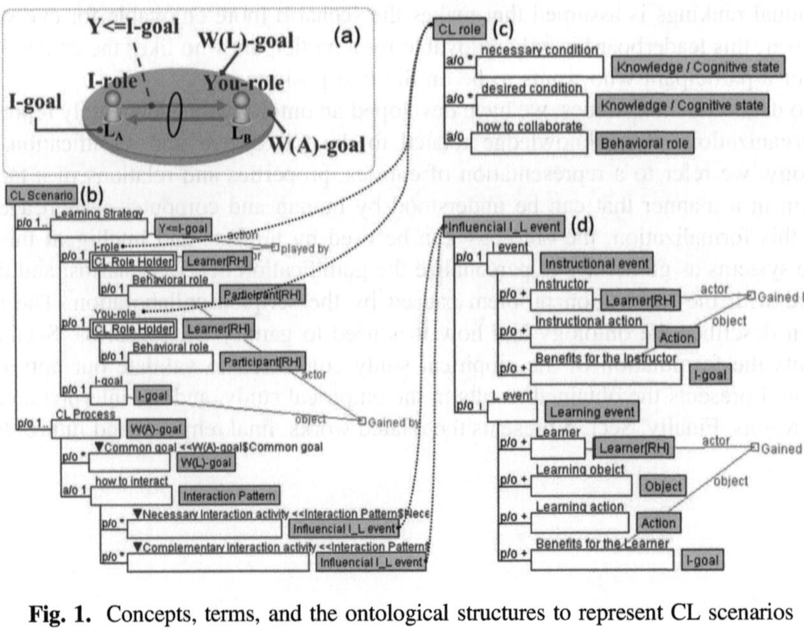

Fig. 1. Concepts, terms, and the ontological structures to represent CL scenarios

To represent CL scenarios, the ontological structure shown in Fig. 1(b) comprises two parts: the learning strategy (*Y <= I-goal*), and the CL process (*W(A)-goal*). The learning strategy describes: the role (*I-role*) for the students in focus "*I*," the role (*You-role*) for the students "*You*" who interact with the student "*I*," and the individual goals (*I-goal*) expected to be achieve for the students "*I*" at the end of CL scenario. The CL process describes the *common goal* (*W(L)-goal*) for the students who participate in the CL scenario and the *Interaction Pattern* used by them to achieve this common goal. This pattern is represented as a sequence of necessary and complementary interactions in which each interaction is defined as an influential Instructional-Learning event (*influential I_L event*) [14]. To represent CL roles, the ontological structure shown in Fig. 1(c) describes the necessary and desired conditions to play a CL role. These conditions refer to the knowledge and/or cognitive states that a student should be have to adequately play the CL role. The representation of influential I_L events as ontological structures is shown in Fig. 1(d) in which an instructional event and a learning event indicate respectively who play the role of instructor and learner in the CL scenario, their instructional and learning actions, the learning object involved in the learning event; the expected *benefits for the instructor*; and the expected *benefits for the learner* (*I-goal*).

Employing the ontological structures shown in Fig. 1, we can represent CL scenarios based on CSCL scripts that are based on instructional/learning theories. For instance, Fig. 2(b–d) shows part of the ontological structures formalized in the ontology to represent CL scenarios that are instantiated from a CSCL script inspired by the Cognitive Apprentice theory shown in Fig. 2(a).

According to the script inspired by the cognitive apprentice theory, for achieving the pedagogical benefits described in this theory, a student who plays the Master role should interact with the students who play the Apprentice role employing the learning strategy "*Learning by Guiding*," whereas the learning strategy "Learning by Apprenticeship" should be employed by the students who play the Apprentice role to interact with the student who plays the Master role. Thus, the ontological structure Cognitive Apprentice shown in Fig. 2(b) comprises: the description of these two learning strategies, and the definition of CL process as Cognitive Apprenticeship type CL session. In this session, the common goal for the learners is *spread of a skill*, and the interaction pattern to achieve this goal is the Cognitive Apprenticeship type Interaction Pattern defined by a set of necessary and complementary interactions enumerated from (1) to (9) in Fig. 2(a).

The Cognitive Apprentice theory indicates that the learners without any knowledge or experience in how to use the target skill should play the role of apprentice in the CL scenarios. This desired condition is shown in Fig. 2(c) in which the knowledge/cognitive states as desired condition to play the role of apprentice are not having the knowledge of how to use the target cognitive skill, not having the knowledge of how to use the target metacognitive skill, not having experience in using the target cognitive skill, and not having experience in using the target metacognitive skill. The interaction "Setting up learning context" is formalized through the ontological structure shown the Fig. 2(d) in which the influential I_L event is defined as the instructional event "Giving

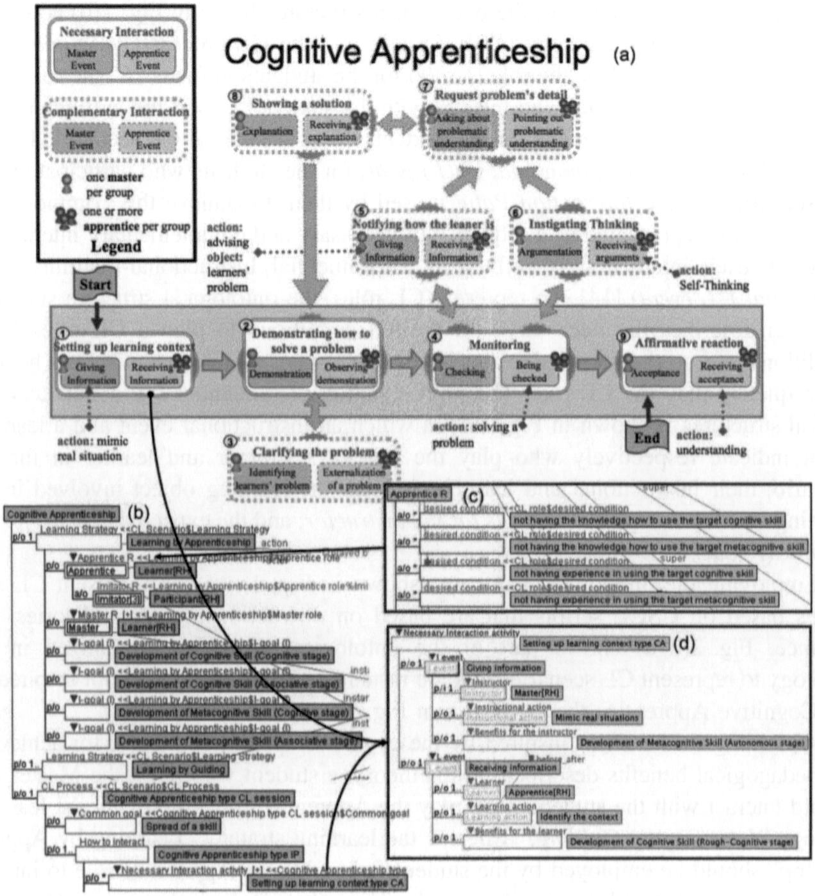

Fig. 2. Part of the ontological structure used to represent cognitive apprentice scenarios.

information" and the instructional event "Receiving information," respectively. Thus, student who plays the master role will attain the development of his metacognitive skill at the autonomous stage performing the action "Mimic real situation," while the students who play the apprentice role will achieve the development of their cognitive skills at the rough-cognitive stage performing the action "Identify the context."

2.2 Ontological Structure to Represent Gamified CL Scenarios

To establish the concepts and semantic relations extracted from the practices and theories related to gamification, and how they are applied in CL scenarios to deal with the motivation problem caused by the scripted collaboration, the concepts and terms shown in Fig. 3 have been defined in the ontology OntoGaCLeS to represent gamified CL scenarios that are compliant with any theory/practice related to gamification.

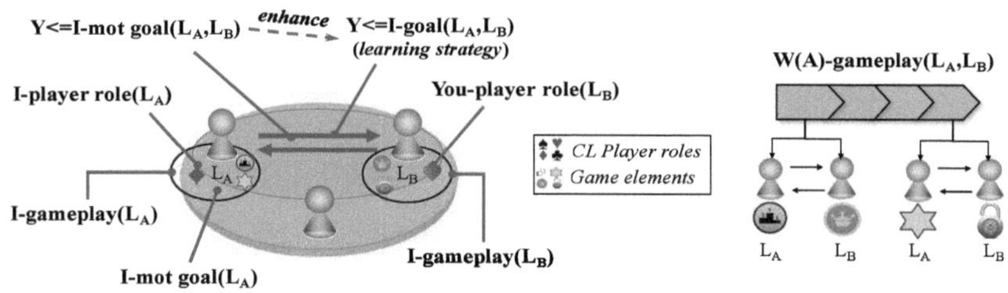

Fig. 3. Concepts and terms defined in the ontology to represent gamified CL scenarios

The concepts and terms (Fig. 3) to represent gamified CL scenarios are formalized by mean of the ontological structure shown in Fig. 4 in which: **Y <= I-mot goal** is the *motivational strategy* to enhance the learning strategy (*Y <= I-goal*) employed by the student in focus "*I*"; **I-mot goal** is the *individual motivational goal* for the student (*I*); **I-player role** and **You-player role** are the *player roles* for the students "*I*" and "*You*", respectively; **I-gameplay** is the *individual gameplay strategy* for the student "*I*"; and **W(A)-gameplay** is the *CL gameplay* for all the students in the CL scenario.

The ontological structure to represent gamified CL scenarios shown in Fig. 4(a) is an extension of structure to represent CL scenarios. In this structure, the motivational strategy (*Y <= I-mot goal*) and the individual gameplay strategy (*I-gameplay*) are introduced in the same level of learning strategy (*I <= goal*); and the CL gameplay (*W (A)-gameplay*) overrides the CL process to indicate: how the students should to interact among them, and how them should interact with the game elements to obtain the benefits of CL process and gamification. The motivational strategy (*Y <= I-mot goal*) also indicates the player role that the student in focus "*I*" should played, the player role for the student (*You*) who interact with "*I*" in the CL process, and the expected benefits (*I-mot goal*) for playing this role. The benefits for playing a player role are described as individual motivational goal (*I-mot goal*) that represent expected changes in the motivational stage of the student "*I*." For describing these goals, currently, two types of individual motivation goals have been formalized in the ontology using need-based theories of motivation as theoretical justification to define the possible stage of students' motivation, the former type is the *satisfaction of psychological needs* and the latter is the as *internalization of motivation.*

For the satisfaction of psychological needs, using the SDT theory [9] and Dan Pink's theory [25] by which an individual is motivated when his/her innate psychological needs are satisfied, we formalized in the ontology the satisfaction of autonomy, competence, mastery, relatedness and purpose as individual motivational goals (*I-mot goal*). The *internalization of motivation* has been formalized in the ontology using the SDT theory as theoretical foundation by which the changes in the motivational stage of an individual are changes from amotivation to extrinsic motivation, and from extrinsic motivation to intrinsic motivation.

The ontological structure to represent player roles is shown in Fig. 4(b). This structure is defined to enable the personalization of gamification based on the classification of students in different player types. The classification of students in different player roles is performed according to the requirements of CL session being gamified,

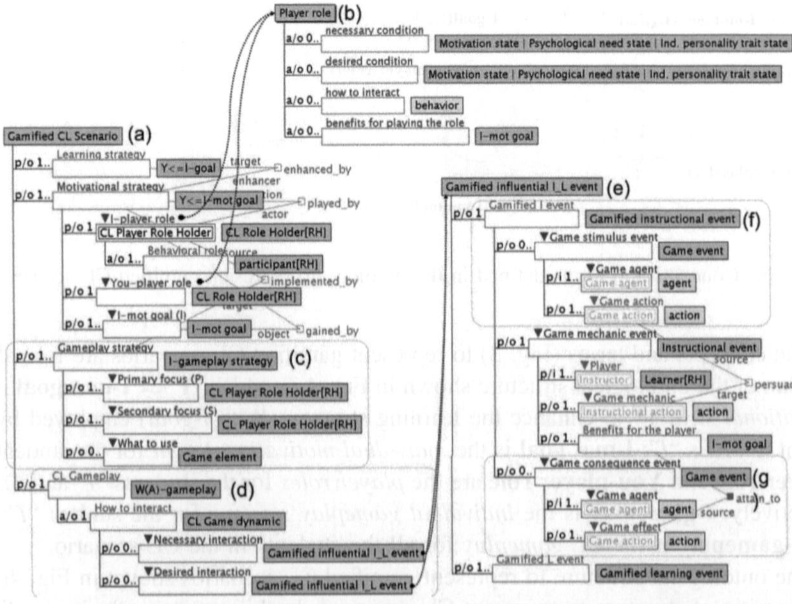

Fig. 4. Ontological structures to represent gamified CL scenarios

and it depends on the necessary and desire condition that are satisfied by the student. These conditions are current psychological states of an individual related to motivation and they are described in various player type models as *psychological needs*, *motivational state*, and *individual personality traits* (e.g. playing styles, cognitive styles). The personalization of gamification based on the player role assigned for a student in focus "*I*" is described in the ontological by mean of the structure "*I-gameplay*" shown in Fig. 4(c). In this structure, the personalization of game elements for the student in focus "*I*" is an arrangement between the player roles assigned for the student in focus "*I*" (*Primary focus*) and "*You*" (*Secondary focus*), and the *game elements* (*What use*) that will be used by the student "*I*" to obtain a motivated and enjoyable experience.

While the individual gameplay strategy (*I-gameplay*) defines the personalization of game elements for the student in focus (I), the CL Gameplay (*W(A)-gameplay*) shown in Fig. 4(d) defines how the game elements provide a motivated and enjoyable experience for all the students in a CL session. Thus, the CL Gameplay is formalized in the ontology as a set of CL game dynamics in which the necessary and complementary interactions between the game elements and students are represented as an extension of *Influential I_L events* named "*Gamified Influential I_L events*" shown in Fig. 4(e). The ontological structure to represent this type of events is composed by two parts: a *Gamified instructional event* and a *Gamified learning event*. In these events, the instructional and learning actions invoked by students who play *player roles* become *game mechanisms* in charge of producing changes in the motivational state of students. These changes are represented as individual motivational goals (*I-mot goal*), and they occur as consequence of actions taken by game agents. These *game actions* (e.g. *promise points* to send an argument) whose function is to *persuade* the students to perform a meaningful learning or instructional action are extracted from the persuasive game design models [24, 30], and they are formalized as a *game stimulus event* shown in Fig. 4(f). As consequence of a meaningful instructional or learning action, in a gamified CL session, the students must receive immediate "feedback" from the game elements (e.g. *move forward one position on a rank* by sending a contra-argument) [2, 17, 33]. Such feedback is formalized as "*game consequence event*" shown in Fig. 4(g) in which *actions* are given by *game agents* to lead the students (instructor or learner) *to attain* his/her individual motivation goals (*I-mot goal*).

2.3 An Ontology-Based Gamification Model for CL Sessions

Employing the ontological structures to represent gamified CL scenarios described in the previous section, it is possible to define gamification models for CL sessions that are compliant with theories and practices related to gamification. In these models, the concepts extracted from different theories and practices are applied for dealing with the motivation problem caused by the scripted collaboration.

Thus, in the ontology, under the assumption that a CL session is not motivating because the students feel/believe that following the interactions indicated by a CSCL script will not lead them to satisfy their current needs, the theories of SDT theory [9] and Dan Pink's theory [25] have been used as theoretical founded to represent the individual motivational goals (*I-mot goal*) as the expected changes that can be produced by the gamification when it is applied as motivational strategy (*Y <= I-mot goal*) in a CL scenario. For instance, in a CL session instantiated from a CSCL Script inspired by the Cognitive Apprentice theory (Fig. 5), the motivational strategy of "Gamifying by Competition and Comparison" (Gamifying by CMPT/CMPR) shown in Fig. 5(a) defines the "*Satisfaction of competence*" and the "*Internalization of motivation*" from the amotivation or extrinsic motivation stage to the intrinsic motivation stage as the individual motivational goals (*I-mot goal*) to be achieved for a student "*I*" who plays the *Master role* when he/she plays the *Achiever role*.

Information extracted from player types models and typology of players has been used to define *Player roles*, as well as, the personalization of game elements defined in the individual gameplay strategies (*I-gameplay strategy*). Figure 5(b) shows the ontological structure formalized to represent the player role "*Achiever role*" based on the Yee's model [32]. In this player roles, the individual preferences over game-components, such as having liking for achievement-components have been used as necessary and desired conditions to represent a player type who likes the components of advancements, mechanics and competition. Employing the description of these components detailed in the Yee's model, we proposed an "individual competition and comparison" gameplay experience (*I-IND CMPT/CMPR gameplay strategy*) shown in Fig. 5(c) as the individual gameplay strategy (*I-gameplay*) that must be experience by a *Master* student who plays the *Yee's Achiever role*. Thus, the game elements used to provide this experience as shown in Fig. 5(c) are a *point system* with *individual* points (*point system (individual)*), a *leaderboard* with an *individual ranking*, an *achievement system* for *participation level*, and a *badge system* for *participation level*.

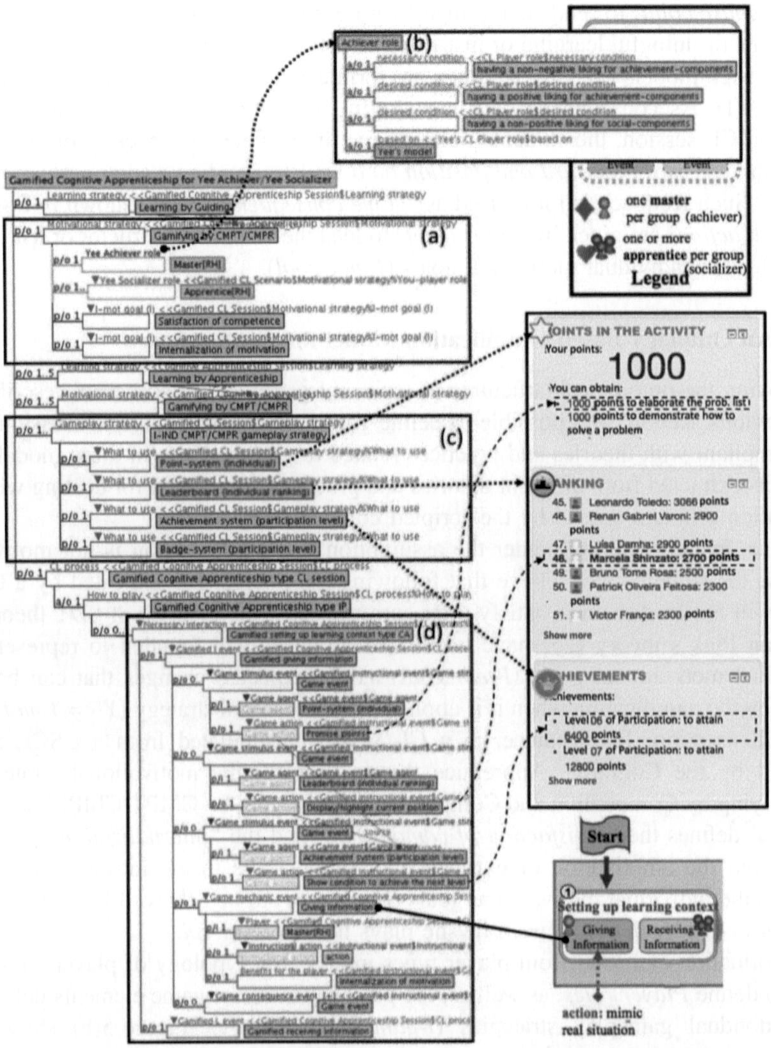

Fig. 5. Part of ontological structure to represent Gamified Cognitive Apprentice scenarios for a master student who plays the achiever role and apprentice students who play socializer role.

As we mentioned in the previous section, persuasive game design models [24, 30] are the information source to define the *Gamified influential I_L events* in gamified CL scenarios. Currently, in the ontology OntoGaCLes, the Persuasive Game Design Strategies (PGDSs) proposed by Orji et al. [24] have been applied in the CL process of CL scenarios based on the Cognitive Apprentice theory, and these PGDSs have been personalized for achievers using the Yee's model. Figure 5(d) shows part of the ontological structure formalized using the PGDSs to represent the event "*Gamified setting up learning context*" in which the instructional event "*Gamified Giving information*" has as stimulus three game events defined for game actions: promise points, display/highlight current position, and show condition to achieve the next level. These game actions are defined to *persuade* the master student to perform the instructional action "*mimic real situation*" defined in the instructional event "*Giving information.*" Furthermore, the gamified I_L event "*Gamified setting up learning context*" defines that the game actions will be respectively done by the game agents: a *point system* with *individual* points, a *leaderboard* with an *individual ranking*, and the *achievement system* for *participation level*. Finally, the expected benefit for the master student is the *internalization of motivation* (*benefits for the player*) as shown in Fig. 5(d).

3 Empirical Study: Validation of Ontology-Based Gamification

Different gamification models for CL sessions had been formalized in the ontology. As shown in the previous section, these models comprise ontological structures that avoid us to gamify a CL session from scratch because they provide enough information to adequately select and personalize the game elements. Such CL sessions in which their game elements are personalized for each student to deal with the motivation problem are known as "*ontology-based gamified CL sessions*" (*ont-gamified CL sessions*), and the empirical study reported here investigates the effects of these sessions on the students' motivation and on the percentage of students by groups who had incomplete participation. To validate the effectiveness of our approach, the effects of the ontology-based CL sessions had been compared with the effects of *non-gamified CL sessions*.

3.1 Research Questions and Hypothesis Formulation

In the empirical study, the first research question $\mathbf{RQ_1}$ is: "whether the ontology-based gamified CL sessions increase the intrinsic motivation of students." Thus, we tested the null hypotheses $\mathbf{H1_{null}}$: "there is no significant difference on the intrinsic motivation of students who participated in ontology-based gamified CL sessions and those who participated in non-gamified CL sessions" against the hypotheses $\mathbf{H1_{alt}}$: "the intrinsic motivation of students who participated in ontology-based gamified CL sessions are greater than those who participated in non-gamified CL sessions." The second research question $\mathbf{RQ_2}$ is: "whether the ontology-based gamified CL sessions have effects in the percentage of students by group who had incomplete participation in the CL session." Thus, we tested the null hypotheses: $\mathbf{H2_{null}}$: "the ontology-based gamified CL sessions have no effect in the percentage of students by groups who had incomplete participation" against the alternative hypothesis $\mathbf{H2_{alt}}$: "the ontology-based gamified CL sessions reduce the percentage of students who had incomplete participation." We define an incomplete participation in the CL session when a student did not perform the necessary interactions described by a script. Thus, to answer the RQ_2, we also tested the null hypotheses $\mathbf{H3_{null}}$: "the ontology-based gamified CL sessions have no effect in the number of meaningful interactions by peers" against the alternative hypotheses $\mathbf{H3_{alt}}$: "the ontology-based gamified CL sessions increase the number of meaningful interactions by peers." In a CL session instantiated from a CSCL script inspired by the

Cognitive Apprentice theory, interaction by peers refers to the pair of interactions between a master student and an apprentice student.

Although it is not the objective of this study to evaluate the effect of ontology-based gamification in the learning outcomes, we need to verify that at least there was no significant decrease in the learning outcomes when the students participated in the ontology-based gamified CL sessions.

3.2 Participants, Design of Empirical Study, and Instruments

In order to achieve the most general results, the context of empirical studies has been a real learning situation in which a CL activity needs the use of CSCL script to achieve a successful learning. Thus, the context selected for the empirical studies was a CL activity of the course "Introduction to Computer Science" taught by Prof. Seiji Isotani at the Institute of Mathematical and Computer Sciences in the University of São Paulo, during the 2nd semester of 2016 (September–December). Therefore, the participants of this empirical study were the 37 undergraduate students signed up in this course. These participants were selected without using a non-probability sampling method, so that the findings found in this evaluation provide lower level of generalization. As these participants are part of a homogeneous population with undergraduate students in the age range from 17 to 25 years old, sharing the same social-economy status and culture, we enclosed the research finding of evaluation in this context. The division of students between control group (student using *non-gamified* CL session) and intervention group (student using *ont-gamified* CL sessions) was randomized. Eighteen students were part of control group and nineteen students were part of intervention group. In reference to the CSCL scripts used to instantiate the CL sessions for the CL activity, a CSCL script based on the Cognitive Apprentice theory [5], and the domain-contents for which this scripts have been instantiated is the subject of "repetition structures in C" (Loops). The empirical study was conducted in three phases during three weeks, one week for each phase, and all these activities were performed through the learning platform Moodle using the modules: Questionnaire (https://goo.gl/J1s8wu), Virtual Programming Lab (VPL) (https://goo.gl/deGtZX), and Scripting-forum (https://goo.gl/HEoYVb).

During the Pre-test Phase (at the First Week). The Questionnaire module was used to collect self-reported information from all students related to their preferences for game components through a survey designed according to the Brazilian-Portuguese Player Profile Questionnaire (QPJ-BR) [1]. During this phase, the students also solved 04 programming problems using the VPL module to evaluate their knowledge and skill of using repetition structures in C. These problems, from the easiest to the most difficult level, are: P1 (Calculate the proper divisors of a number), P2 (Calculate the sum of prime divisors of number), P3 (Calculate distance of rebounds for an elastic ball), and P4 (Calculate the maximum length of a hailstone sequence).

During the Post-test Phase (at the Third Week). The students answered the questions of the Intrinsic Motivation Inventory (IMI) [21] through the questionnaire module. In this phase, it was also requested for the students to solve 04 programming problems in the environment of VPL module. These problems related to the topic of repetition structures in C (from the easiest to the most difficult level) are: PA (Calculate the Inverse Fibonacci sequence on base *n* and *m*), PB (Calculate the absolute difference between odd and even numbers in an inverse Fibonacci sequence), PC (Calculate the i-th prize of a machine slot), and PD (Calculate the highest prize of a machine slot).

During the Execution Phase (at the Second Week). The students participated in a CL activity to improve their knowledge and skills of using repetition structure in C. To perform this activity, the students were grouped together into groups of four and five students to participate in either a non-gamified (students of control group) or ontology-based gamified CL session (students of intervention group). During these sessions, the scripting-forum module was used to structure the collaboration and discussion according to a script inspired by the Cognitive Apprenticeship theory shown in Fig. 2(a). Students who solved the problem P3 or P4 during the pre-test phase played the master role during the CL session, and students who did not solved these problems played the apprentice role.

The script shown in Fig. 2(a) was used in this empirical study to structure the collaboration among the students during the CL sessions in which the interaction to be performed by the students consists in: (1) *Setting up the learning context* - the student who played the master role mimic a teacher requesting to elaborate a list of at least two exercises related to the subject "repetition structure in C" for the students who play the apprentice role; (2) *Demonstrating how to solve a problem* - the master students present for the apprentices an explanation about how to solved a problem related to the subject "repetition structure in C;" (3) *Clarifying the problem* - after demonstrating how to solve a problem, if some apprentice has doubts about the explanation presented by the master, the master student identify and clarify the apprentice's doubts; (4) *Monitoring* - the apprentices solve and send their solutions for the list of problems. During the interaction "*monitoring*," the master student checks the solution sent by the apprentice, and if this solution is correct, the master gives acceptance to the solution performing the interactions (9) *Affirmative reaction*. If the solution is wrong, the master gives an advice about how to solve the problem by performing the interaction (5) *Notifying how the learner is*. If the solution has errors identified, the *master* indicates the errors in the solution by performing the interaction (6) *Instigating thinking*. If an apprentice cannot solve a problem despite the help of the master student, he should indicate his difficulty to solve a problem performing the interaction: (7) *Request problem's detail*, and (8) *Shows the solution for the problem*.

The CL sessions employed for the intervention group were gamified employing the information provided by an ontology-based gamification model that defines two types of Gamified Cognitive Apprenticeship scenarios: one scenario for students who play the achievers role (*Gamified Cognitive Apprenticeship Session for Achievements*); and one scenario for students who play the socializer role (*Gamified Cognitive Apprenticeship Session for Socializers*). To establish these player roles in the intervention group, we used algorithm proposed in [4] in which the information of students collected through the QPJ-BR were compared with the preferences for game components provided by the ontological structure to represent Player roles. Thus, students who obtained higher score on their preference for social-component were grouped to work together in a Gamified Cognitive Apprenticeship Session for Socializers playing the Socializer role; whereas students who obtained higher score on their preference for achievement-components were grouped to work in a Gamified Cognitive Apprenticeship Session for Achievements playing the Achiever role.

The ontological structures defined in the previous section to represent Gamified CL Scenarios were used as guidelines to configure the personalization of game elements in the Gamified Cognitive Apprenticeship Sessions. According to these guidelines, the Gamified Cognitive Apprenticeship Sessions for Achievers were set-up to provide an "individual competition and comparison" gameplay experience (*I-IND CMPT/CMPR gameplay strategy*) for the participants, whereas the Gamified Cognitive Apprenticeship Sessions for Achievers were set-up to provide a "cooperative competition and comparison" gameplay experience (*I-COOP CMPT/CMPR gameplay strategy*). Figure 5 shows the ontological structure used to configure the game elements defined in the Gamified Cognitive Apprenticeship Sessions for Achievers, where: the point system had been set-up as a point system that only indicates the points obtained for each student of CL session, the ranking system had been defined as an individual ranking of participants, and the badges system had been configure to highlight the individual achievements obtained by the students. In the Gamified Cognitive Apprenticeship Sessions for Socializers, the point system had been set-up as group point system that indicated the points obtained for the groups during the CL activity, the ranking system had been defined as a ranking of groups, and the badges system had been configure to highlight the group achievements obtained by the students.

The complete information of CL session and the materials employed by the students during this empirical study are available at the site: https://goo.gl/tFTxjB.

4 Analysis Results of Empirical Study and Hypotheses Testing

Prior to perform the statistical analysis related to the intrinsic motivation, the collected data from 32 students[1] were evaluated to identified careless responses. A careless response is defined as a response in which more than half of successive items have the same value. This response correspond to students who intentionally or unintentionally answered the IMI questionnaire in a manner that does not accurately reflect their true beliefs or feelings. Thus, two careless response were identified and removed from the analysis related to intrinsic motivation, so that the data analysis was performed using N = 30 participants. An exploratory factor analysis and a reliability test (available at the website: https://goo.gl/pZHtdN) were conducted to validate the Portuguese version of IMI questionnaire employed in the empirical. According to this analysis, the cumulative variances (>0.60) and proportion explained (47% for the first component) obtained in the factorial analysis, the IMI questionnaire is consistent with the psychometric validation of original IMI questionnaire [21], and according to the reliability test, this IMI questionnaire had an excellent consistency of 0.90 in the Cronbach's alpha (α).

[1] Although the participants were 37 students, only 32 students answered the IMI questionnaire during the post-test because answering the IMI questionnaire was not a mandatory activity.

In order to answer the research question **RQ_1**:"whether the ontology-based gamified CL sessions increase the intrinsic motivation of students," the Mann-Whitney's U test [22] was employed as the statistical method to find significant differences in the intrinsic motivation of students who participated in either ontology-based gamified or non-gamified Cognitive Apprentice sessions. A non-parametric method was chosen instead the parametric t-test because the normality assumption was rejected in the variable Pressure/Tension, and the number of subjects in each group is less than 20, so that a parametric test cannot be performed with skewed and non-normal distributions as is indicated in [3, 10]. Table 1 shows the descriptive statistic results obtained for this test. According to these results, there is a significant difference in the overall intrinsic motivation with values of U = 167.5, Z = 2.309, p = 0.0099 and r = 0.42 (*medium effect size*). Here, we reject the null hypothesis **$H1_{null}$**: "There is no difference on the intrinsic motivation of students who participate in ontology-based gamified CL sessions and those who participate in non-gamified CL sessions." According to this study, students who participated in ontology-based gamified CL sessions (*median = 4.80*) reported to be more intrinsic motivation than students who participated in non-gamified CL sessions (*median = 3.58*). Therefore, this case study becomes an evidence to support the alternative hypothesis **$H1_{alt}$**: "The intrinsic motivation of students who participate in ontology-based gamified CL sessions are greater than those who participate in non-gamified CL sessions."

Table 1. Mann-Whitney's U test results of the effect on the intrinsic motivation of students

	Group	N	Median	Mean ranks	Sum ranks	U	Z	p-value	r
Interest/enjoyment	ont-gamified	16	4.78	18.84	301.5	165.5	2.228	0.0124	0.40
	non-gamified	14	2.57	11.67	163.5				
Perceived choice	ont-gamified	16	4.70	18.31	293.0	157.0	1.873	0.0308	0.34
	non-gamified	14	4.00	12.28	172.0				
Pressure/tension	ont-gamified	16	2.50	12.56	201.0	65.0	−1.961	0.0250	0.35
	non-gamified	14	3.66	18.85	264.0				
Effort/importance	ont-gamified	16	5.00	15.90	254.5	118.5	0.271	0.3985	0.04
	non-gamified	14	5.16	15.03	210.5				
Intrinsic motivation	ont-gamified	16	4.80	18.96	303.5	167.5	2.309	0.0099	0.42
	non-gamified	14	3.58	11.53	161.5				

Students who participated in ontology-based gamified CL sessions reported higher significant scores for the interest/enjoyment (*median = 4.78*) and perceived choice (*median = 4.70*) than students who participated in non-gamified CL sessions (*median = 2.57* for interest/enjoyment; and *median = 4.00* for perceived choice). Students who participated in ontology-based CL sessions also reported lower significant scores for the pressure/tension (*median = 2.50*) than students who participated in non-gamified CL sessions (*median = 3.66*). Therefore, we can state that the gamified CL sessions obtained by our approach produce positive results in the intrinsic motivation of students towards the reduction of the pressure/tension and by increasing the interest/enjoyment and perceived choice.

To answer the research question **RQ₂**: "whether the ontology-based gamified CL sessions have effects in the percentage of students by group who had incomplete participation in the CL session," we performed the Mann-Whitney's U test in the percentage of students who had incomplete participation and in the number of interactions by peers. In both cases, the data follow a non-normal distribution, and the number of participants were less than 20 in each group, so that a non-parametric method should be used instead to a parametric method [3, 10]. Figure 6 shows the results of Mann-Whitney's U test in which there is a significant difference for the percentage of incomplete participation with values of $U = 4$, $Z = -1.79$, p-value = 0.04 and $r = 0.56$ (*medium size*). The percentage of students by group who had incomplete participation in ontology-based gamified CL session (*median = 0.25*) was lower than the percentage of students who had incomplete participation in non-gamified CL sessions (*median = 0.6*) – Fig. 6(a). Thus, we reject the null hypothesis $\mathbf{H2_{null}}$: "the ontology-based gamified CL sessions have no effect in the percentage of students by group who had incomplete participation," so that this case study becomes an evidence in favor of alternative hypothesis $\mathbf{H2_{alt}}$: "the ontology-based gamified CL sessions reduce the percentage of students who had incomplete participation." Although, the media related to the number of interactions by peers in gamified CL sessions (*median = 4*) was greater than the media obtained for non-gamified CL sessions (*median = 3*) – Fig. 6(b), we did not find significant difference, so that we cannot refute the null hypothesis $\mathbf{H3_{null}}$: "the ontology-based gamified CL sessions have no effect in the number of meaningful interactions by peers."

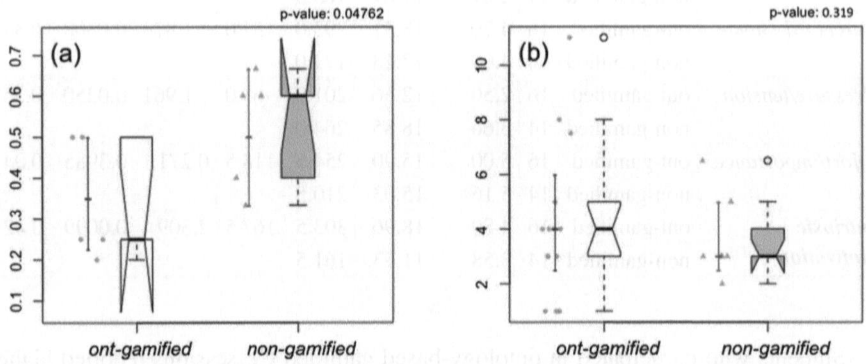

Fig. 6. (a) Percentage of students by group who had incomplete participation, and (b) number of interactions by peers

In relation to the learning outcomes, we tested the null hypotheses $\mathbf{H4_{null}}$: "the ontology-based gamified CL sessions have no effect in the correctness score and the number of submissions to solve programming problems" against the hypotheses $\mathbf{H4_{alt}}$:

"the ontology-based gamified CL sessions have effect in the correctness score and the number of submissions to solve programming problems." After running the Mann-Whitney's U test, we did not find significant differences in relation to the correctness score and the number of submission to solve programming problems.

5 Related Works, Final Remarks and Future Steps

Although several studies have shown that the students' motivation plays an important role in CL process [7, 15, 27], and gamification has proven to be a technology to successfully motivate and engage people in different situations [28, 29], to the best knowledge of the authors of this chapter, no one attempt has been made to gamify CL scenarios, and thus to deal with the motivation problem when CSCL scripts are used as method to structure and orchestrate the CL process. Currently, there are research works in which self-regulation learning and emotion regulation are proposed as method to deal with the motivation problem caused by the scripted collaboration [8, 16]. However, these solutions only evoke the intrinsic motivation of participants, and students who are not interesting in the domain-content to be learn in the CL process are not motivated. The gamification uses intrinsic and extrinsic motivators, so that this technology could be used to engage and motivate the participants who are interesting in the domain-content and the participants who are not interesting in the content domain. Thus, the gamification of CL scenarios looks like a promising solution to deal with the motivation problem caused by the scripted collaboration. But firstly, it is necessary to provide a way to adequately represent the well-thought-out design of gamified CL scenarios.

In this sense, we proposed the use of ontologies to represent the knowledge extracted from the best practices and theories related to gamification. Ontological structures has been defined to represent gamified CL scenarios that are compliant with theories and practices related to gamification (Sect. 2.2), and these structures aim to provide structured guidance on how to gamify CL sessions through ontology-based models of gamification (as was shown in Sect. 2.3). To validate this approach in which ontologies and gamification are used to deal with the motivation problem caused by the scripted collaboration, we conducted an empirical study in a real scenario in which we analyzed and compared the effects of ontology-based gamified CL sessions and non-gamified CL sessions on the students' motivation and the percentage of students by group who has incomplete participation. Our approach of using ontologies and gamification significantly increased the intrinsic motivation of students, and it also significantly reduced the percentage of students by group who had incomplete participation in CL sessions. These results are evidence to demonstrate the effectiveness of our approach for dealing with the motivation problem caused by the scripted collaboration in the context of Brazilian computer science undergraduate students in the 17–25 age range.

In the current version of our ontological structures, we did not formalize concepts extracted from the flow theory, player's journey, and fun for game design that are considered relevant to deal with the motivation problem caused by the scripted collaboration. Thus, our next steps will consider how to formalize these concepts in the gamification of CL scenarios. Furthermore, for future researches, we will extend the evaluation presented in this study to evaluate the efficiency of our approach by comparing the effects of the ontology-based CL sessions and the effects of the CL sessions that have been gamified without using the support given by the ontologies. We also will analyze the effects of our approach on the learning outcomes, and the correlation of these effects with the effects on the students' motivation.

Acknowledgements. This work was partially funded by CNPq, CAPES and FAPESP.

References

1. Andrade, F., et al.: QPJ-BR: Questionário para Identificação de Perfis de Jogadores para o Português-Brasileiro. In: XXVII Brazilian Symposium on Computers in Education, Maceio, Brazil, pp. 637–646 (2016)
2. Bleumers, L., et al.: State of play of digital games for empowerment and inclusion: a review of the literature and empirical cases. Eur. Com. **10**, 36295 (2012)
3. Briand, L., et al.: On the application of measurement theory in software engineering. Empir. Softw. Eng. **1**(1), 61–88 (1996)
4. Challco, G.C., Moreira, D.A., Mizoguchi, R., Isotani, S.: An ontology engineering approach to gamify collaborative learning scenarios. In: Baloian, N., Burstein, F., Ogata, H., Santoro, F., Zurita, G. (eds.) CRIWG 2014. LNCS, vol. 8658, pp. 185–198. Springer, Cham (2014). https://doi.org/10.1007/978-3-319-10166-8_17
5. Collins, A.: Cognitive apprenticeship and instructional technology. In: Educational Values and Cognitive Instruction: Implications for Reform, pp. 121–138 (1991)
6. Deterding, S., et al.: From game design elements to gamefulness: defining gamification. In: Proceedings of the 15th International Academic MindTrek Conference: Envisioning Future Media Environments, pp. 9–15. ACM (2011)
7. Dillenbourg, P.: Over-scripting CSCL: the risks of blending collaborative learning with instructional design. In: Kirschner, P.A. (ed.) Three Worlds of CSCL. Can We Support CSCL? pp. 61–91. Open Universiteit Nederland, Nederland (2002)
8. Feidakis, M., et al.: Providing emotion awareness and affective feedback to virtualised collaborative learning scenarios. Int. J. Contin. Eng. Educ. Life Long Learn. **24**(2), 141–167 (2014)
9. Gagné, M., Deci, E.L.: Self-determination theory and work motivation. J. Organ. Behav. **26** (4), 331–362 (2005)
10. Guadagnoli, E., Velicer, W.F.: Relation to sample size to the stability of component patterns. Psychol. Bull. **103**(2), 265–275 (1988)
11. Inaba, A., et al.: An interaction analysis support system for CSCL: an ontological approach to support instructional design process. In: 2002 Proceedings of the International Conference on Computers in Education, vol. 1, pp. 358–362 (2002)
12. Inaba, A., et al.: What learning patterns are effective for a learner's growth. In: Proceedings of the International Conference on Artificial Intelligence in Education, Sydney, pp. 219–226 (2003)
13. Isotani, S., Mizoguchi, R., Isotani, S., Capeli, O.M., Isotani, N., de Albuquerque, A.R.P.L.: An authoring tool to support the design and use of theory-based collaborative learning activities. In: Aleven, V., Kay, J., Mostow, J. (eds.) ITS 2010. LNCS, vol. 6095, pp. 92–102. Springer, Heidelberg (2010). https://doi.org/10.1007/978-3-642-13437-1_10

14. Isotani, S., et al.: An ontology engineering approach to the realization of theory-driven group formation. Int. J. Comput.-Support. Collab. Learn. **4**(4), 445–478 (2009)
15. Järvelä, S., et al.: Research on motivation in collaborative learning: moving beyond the cognitive-situative divide and combining individual and social processes. Educ. Psychol. **45** (1), 15–27 (2010)
16. Järvelä, S., et al.: Self-regulation and motivation in computer-supported collaborative learning environments. In: Learning Across Sites: New Tools, Infrastructures and Practices, pp. 330–345 (2011)
17. Kapp, K.M.: The Gamification of Learning and Instruction: Game-Based Methods and Strategies for Training and Education. Pfeiffer, San Francisco (2012)
18. King, A.: Scripting collaborative learning processes: a cognitive perspective. In: Fischer, F., Kollar, I., Mandl, H., Haake, J.M. (eds.) Scripting Computer-Supported Collaborative Learning. CULS, vol. 6, pp. 13–37. Springer, Boston (2007). https://doi.org/10.1007/978-0-387-36949-5_2
19. Knutas, A., et al.: Increasing collaborative communications in a programming course with gamification: a case study. In: Proceedings of the 15th International Conference on Computer Systems and Technologies, pp. 370–377. ACM, New York (2014)
20. Kong, J.S.-L., et al.: The effects of peer intrinsic and extrinsic motivation on MMOG game-based collaborative learning. Inf. Manag. **49**(1), 1–9 (2012)
21. McAuley, E., et al.: Psychometric properties of the intrinsic motivation inventory in a competitive sport setting: a confirmatory factor analysis. Res. Q. Exerc. Sport **60**(1), 48–58 (1989)
22. McKnight, P.E., Najab, J.: Mann-Whitney U test. In: Corsini Encyclopedia of Psychology. Wiley (2010)
23. Mizoguchi, R., et al.: The model of roles within an ontology development tool: Hozo. Appl. Ontol. Roles Interdisc. Perspect. **2**(2), 159–179 (2007)
24. Orji, R., et al.: Modeling the efficacy of persuasive strategies for different gamer types in serious games for health. User Model. User-Adapt. Interact. **24**(5), 453–498 (2014)
25. Pink, D.H.: Drive: The Surprising Truth About What Motivates Us. Riverhead Books, New York (2011)
26. Prieto, L.P., et al.: Exploring teachers' perceptions on different CSCL script editing tools. Comput. Educ. **78**(Supplement C), 383–396 (2014)
27. Rienties, B., et al.: The role of academic motivation in computer-supported collaborative learning. Comput. Hum. Behav. **25**(6), 1195–1206 (2009)
28. Seaborn, K., Fels, D.I.: Gamification in theory and action: a survey. Int. J. Hum.-Comput. Stud. **74**, 14–31 (2015)
29. de Sousa Borges, S., et al.: A systematic mapping on gamification applied to education. In: Proceedings of the 29th Annual ACM Symposium on Applied Computing, pp. 216–222. ACM (2014)
30. Torning, K.: A review of four persuasive design models. Int. J. Concept. Struct. Smart Appl. **1**(2), 17–27 (2013)
31. Webb, E.N.: Gamification: when it works, when it doesn't. In: Marcus, A. (ed.) DUXU 2013. LNCS, vol. 8013, pp. 608–614. Springer, Heidelberg (2013). https://doi.org/10.1007/978-3-642-39241-2_67
32. Yee, N.: Motivations for play in online games. Cyberpsychol. Behav. **9**(6), 772–775 (2006)
33. Zichermann, G.: Fun is the future: mastering gamification (2010)

An Agile Method for Developing OERs and Its Application in Serious Game Design

Rafaela V. Rocha[1,3](✉), Pedro H. D. Valle[1], José C. Maldonado[1], Ig Ibert Bittencourt[2], and Seiji Isotani[1]

[1] Institute of Mathematics and Computer Science, University of São Paulo, São Carlos, Brazil
rafaela.vilela@gmail.com, pedrohdvalle@gmail.com, {jcmaldon,sisotani}@icmc.usp.br
[2] Computing Institute, Federal University of Alagoas, Maceió, Brazil
ig.ibert@ic.ufal.br
[3] Center of Mathematics, Computing and Cognition, Federal University of ABC, Santo André, Brazil

Abstract. Interactive educational resources can promote a virtual environment that enable their users to acquire knowledge, train skills and have constant feedback. However, for an educational resource to be effective, its development should be systematic and multidisciplinary. Different methods have been proposed to provide support for the educational resources development. However, they do not describe the entire lifecycle, including the multiple requirements, such as, learning objectives and content definition, learner performance assessment, nonlinearity, student engagement and continuous feedback. This paper describes an overview of the AIMED method, an agile, integrative and open method for educational resources development, in a holistic and multidisciplinary way. It integrates the areas of simulation modelling, game design, instructional design, principles of agile development methods, project management and content domain. AIMED is based on integration of DevJSTA methodology, for creation of serious games, and AM-OER method, for development of open educational resources. A feasibility study was done with one serious game development.

Keywords: Agile method · Educational resource · Serious games

1 Introduction

In educational resources, such as, serious games and training simulations, students acquire competences that can be transferred to the real world [1–5]. They offer a regular and diversified content that can be used to prepare a greater number of students. However, to be successful as a final product, their development should be systematic and multidisciplinary, and often, this is complex and costly.

Different methods have been proposed to provide support for the serious games development, [6–19], in addition to specific methodologies in simulation [20–24], game [25–27], and learning and training [28–31]. In general, they do not describe the

A. I. Cristea et al. (Eds.): HEFA 2017, CCIS 832, pp. 192–206, 2018.
https://doi.org/10.1007/978-3-319-97934-2_12

entire lifecycle, including the multiple requirements of serious games: learning objectives and content definition, learner performance evaluation, logical and physical simulation fidelity, nonlinearity, student engagement with the inclusion of challenges, rewards, levels and continuous feedback [10, 32]. Methodologies for game design, simulation modelling, or instructional design aim only the game elements, or simulation, or instruction. Methodologies for development of serious games and educational simulations integrate some elements, but do not include all the necessary requirements, creating a lack of balance in the final product; for example, an educational game with adequate content, but that does not motivate and engage students, or a playful game, that does not provide the intended learning [13, 32–36].

The main objective of this paper is to propose and describe an Agile, Integrative and open Method for open Educational resources Development (AIMED), to create serious games and similar educational resources used for learning and evaluation. To achieve our goals, we first review the literature and related works. After, AIMED was proposed and specified, in a holistic and multidisciplinary approach, based on DevJ-STA (for serious games [19]) and AM-OER (for open educational resources [12]). Finally, we implemented and described its use for a serious game production.

This paper is organized as follows. Literature review are described in Sect. 2. AIMED method is presented in Sect. 3. Its use is related in Sect. 4. Discussion is described in Sect. 5, followed by conclusions and future works in Sect. 6.

2 Literature Review

There are different resources that can be used in education, as well, there are different methods to create them. For example, entertainment games [25–27], simulations [20–24], learning objects [28–30] and training materials [31] have specific methods. However, there are significant differences among educational games, interactive simulations, learning objects and software. These differences are in main objectives (entertainment, learning and assessment), reality representation (not real, abstract, real or verisimilitude), content linearity (linear or not) and focus (gamer, user, trainer, student - who plays different roles).

Other methods are combination of these or extension of traditional software development lifecycle. An example is the agile AM-OER method, that aims to develop Open Educational Resources (OER - including educational games) [11]. It is based on software engineering practices (XP [37], SCRUM [38] and Agile design practices [39]) and learning design [40], to improve the quality of the OER and to facilitate its adaptation and reuse [19]. AM-OER consists of three phases with these processes: (1) Kickoff: initial architecture; (2) Development: sprint planning, iterative design, incremental and iterative development, and sprint assessment; (3) Sharing: delivery of the modules. The AM-OER benefits include incremental development with continuous delivery, being an agile, flexible to change and iterative method. However, it does not specify the learning assessment, game design and simulation modelling.

The focus of this project is the development of interactive educational resources, such as, educational games, serious games, interactive simulations, gamified learning objects, and virtual environments for educational and training purposes. Therefore, in the next subsections, methods for developing educational games (2.1) and simulations and serious games (2.2) are presented.

2.1 Methodologies for Educational Simulation and Game Development

Educational games do not simulate system and process behaviours; they aim to provide and evaluate knowledge [10]. On the other hand, educational simulations simulate reality to some degree (such as, physics simulators), but do not include game elements that can engage and challenge the learners.

The methodologies of simulation and educational game development are: Digital Educational Game Development Methodology (GAMED) [8], Fuzzified Instructional Design Development of Game-like Environments for learning (FIDGE) [9], Systematic Process [10] and Digital Game Based Learning-Instructional Design Model (DGBL-ID) [11]. In general, they do not address the logical and physical fidelity requirements of the simulation, do not provide support for student assessment and feedback, and do not integrate effectively gameplay and instruction.

2.2 Methodologies for Serious Games Development

The methodologies of serious game development are: Synergy Process [13], Development Process for Serious Games [14], Simulation-Games Instructional Systems Design Model (SG-ISD) [15], Serious Game Unified Process (SGUP) [16] and Framework for Conceptual Modelling for Simulation-based Serious Gaming [17]. They focus on instructional and game design, but without detailed integration. They do not specify the artefacts developed and the activities that should be performed by each team member. They do not support the serious game evaluation and student's performance assessments. In addition, some approach only a few phases of development [14, 19].

An iterative and integrative methodology was proposed by Rocha [19], named DevJSTA, which is based on game design practices, simulations modelling, pedagogical design. Its main objective is to integrate game, content, simulation, training of required competences, assessment, feedback, evaluations and validations to support the production, reuse, interoperability and extension of new serious games. It consists of three phases with the processes: (1) Pre-production: planning; (2) Production: analysis, design, implementation, integration and testing; And (3) Post-production: execution and results evaluation. In addition, there is the verification and validation process that covers the entire lifecycle.

The DevJSTA benefits include the integration of different areas to develop serious games (i.e., different professionals need to collaborate to produce the integrating artefacts of each process), and the inclusion of different evaluations (both serious game developed process and training done with it) [19]. However, it is a methodology based on cascade development, that is, although iterative, there is a task prioritization activity, incremental development and continuous delivery. Thus, the methodology is also not flexible in aspects of changes during the development or for production of other resources that do not integrate game, simulation, content and assessment. To overcome the current limitations, we propose an agile and integrative method, named AIMED, described in the next section.

3 Agile, Integrative and Open Method for Open Educational Resources Development (AIMED)

AIMED is an agile, integrative and open method for open educational resources development, that integrates practices of instructional design, game design, simulation modelling, software engineering and project management; to support the development of efficient and effective educational resources (mainly Open Educational Resources - OERs) [33]. These educational resources are primarily educational games, serious games, interactive simulations, gamified learning objects, and virtual environments for educational and training purposes. AIMED is mainly an evolution and integration of the DevJSTA methodology for the development of serious games [19] and the Agile Method for the development of Open Educational Resources (AM-OER) [12].

To overcome the limitations of DevJSTA and AM-OER, the principles of agile development methods and project management practices were included. These principles are: (1) Individuals and interactions: collaborative and integrated participation of different professionals (people with different and complementary skills and roles); (2) Working educational resources: incremental development with continuous delivery; (3) Customer collaboration: collaborative participation of the customer, with continuous evaluation of the deliverable; and (4) Responding to change: flexible and iterative lifecycle. Project management practices aim to add planning and controlling of the resources (financial cost, human, and time) used during the development - activities that are not covered by DevJSTA and AM-OER.

The AIMED method consists of five macroprocesses with fourteen processes (that are described in the next subsections): (1) Organizational; (2) Pre-Production; (3) Production; (4) Post-production; and (5) Support Processes; as showed in Fig. 1. Artefacts were proposed for each phase (vision and initial plans, specification models, project models and diagrams).

Each process has different groups of activities that guide its development. Depending on the type of educational resource, these groups may or may not occur, and even in parallel with others. Workflows were specified using Business Process Model and Notation (BPMN). Thus, the notation elements called "gateways" possible describe what is required or optional, based on conditions, to enable the mapping of the decisions taken during the development.

The AIMED method is based on agile methods for the rapid and continuous production of deliverables artefacts, that can be designed, produced and evaluated in each iteration. The processes are described in the following subsections.

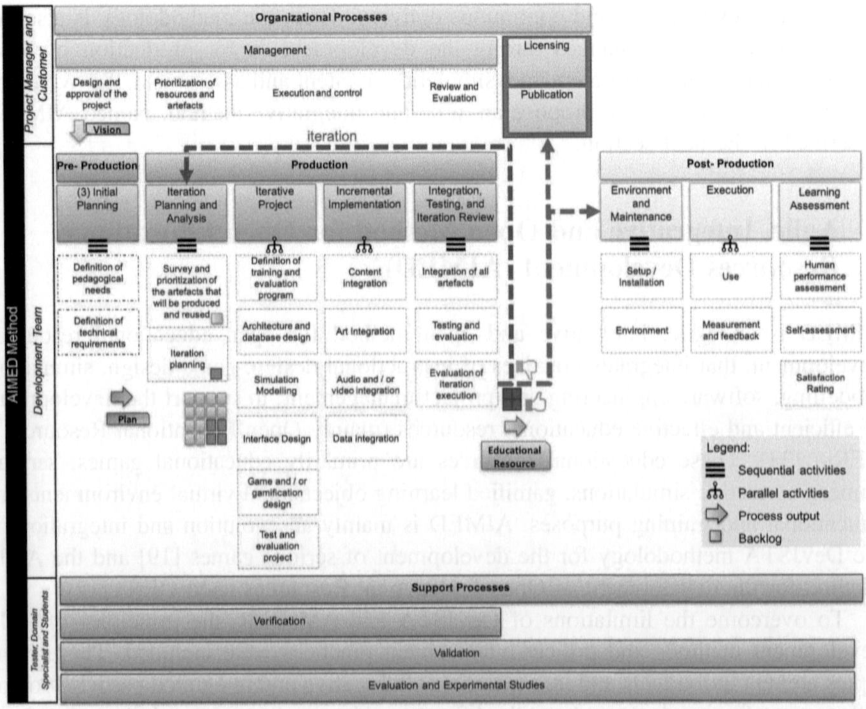

Fig. 1. Overview of AIMED method: processes and activities.

3.1 Organizational Processes

There are three organizational processes [33]:

(1) Management: performed throughout pre-production and production. It contains activities to: (1.1) Design and approval of the project (presented in Fig. 2); (1.2) Prioritization of resources and artefacts; (1.3) Execution and control; and (1.4) Review and evaluation. The project manager is responsible for managing and controlling the artefacts and the project (human resources, budget and deadlines). The costumer is responsible for approving and financing the project.

(2) Licensing: contains activities and tasks to define licensing policies and metadata (2.1 Definition of licensing policies and 2.2 Definition of sharing metadata). The costumer defines copyrights and budget of the educational resource. The project manager controls and describe license and metadata.

(3) Publication: contains the activities: (3.1) Production of tutorials (pedagogical, access, user and development) and (3.2) Sharing in repositories. Licensing and Publication are executed at the end of production process.

3.2 Pre-production Process

In the pre-production, there is the:

(4) Initial Planning process, that contains the activities presented in Fig. 2: (4.1) Definition of pedagogical requirements (learning, training and assessment); and (4.2) Definition of technical requirements (game, simulation, software and interface). The educator is responsible for the pedagogical aspects, and Development Team for the technical aspects [33].

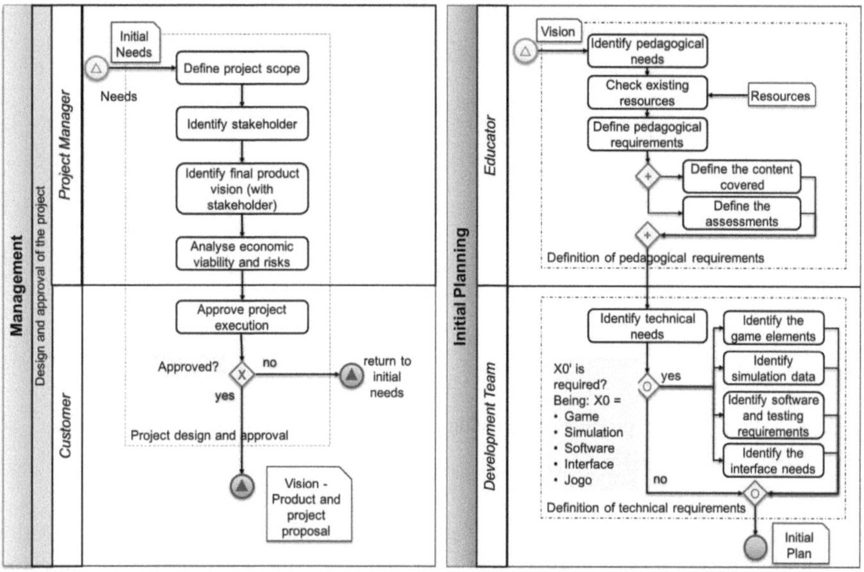

Fig. 2. Overview of the activities: design and approval of the project (management process) and definition of pedagogical and technical requirements (initial planning).

3.3 Production Processes

This macroprocess contains four iterative processes (Figs. 3 and 4) [33]:

(5) Analysis and Planning of the Iteration: contains activities to identify and plan the artefacts that will be produced at each iteration (this includes the reuse, review and remix). (5.1 Identification and prioritization of artefacts that will be produced and reused and 5.2 Iteration planning.) The educator is responsible for identifying these required artefacts and Development Team and Project Manager are responsible for selecting and including them in each iteration.

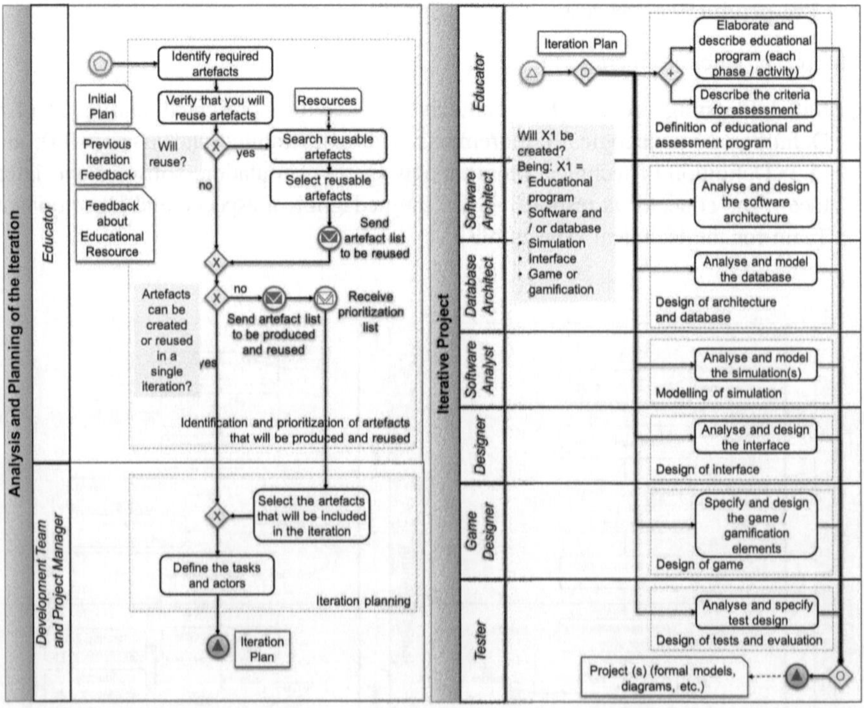

Fig. 3. Overview of the production activities: "analysis and planning of the iteration" and "iterative project" processes.

(6) Iterative Project: consists of the following activities that can be performed in parallel: (6.1) Definition of educational and assessment program (by Educator); (6.2) Architecture and database design (by Software and Database Architects); (6.3) Simulation modelling (by Software Analyst); (6.4) Interface design (by designer); (6.5) Game or gamification design (game designer); (6.6) Design of tests and evaluation (by tester).

(7) Incremental Implementation: consists of creating, reviewing, reusing, and remixing artefacts. (7.1) Content integration; (7.2) Art integration; (7.3) Audio and/or video integration; (7.4) Data integration. Author is responsible for creating, reviewing, reusing and remixing content (texts, tables, charts, diagrams, questionnaires, databases, videos, etc.). Artist, Designer and/or 3D Animator is responsible for art resources (images, textures, photographs, comics, 3D objects, 2D and 3D animations, graphic effects, etc.). Musician and/or Audio Designer is responsible for audio resources (sounds, voices, music, special effects, etc.). Writer and/or Video Editor is responsible for video resources (videos, videotapes, videoconferences, online seminars, etc.). Programmer is responsible for programming, reviewing and reusing codes (components, classes, modules, etc.). Database administrator is responsible for database (tables, scripts, queries, etc.).

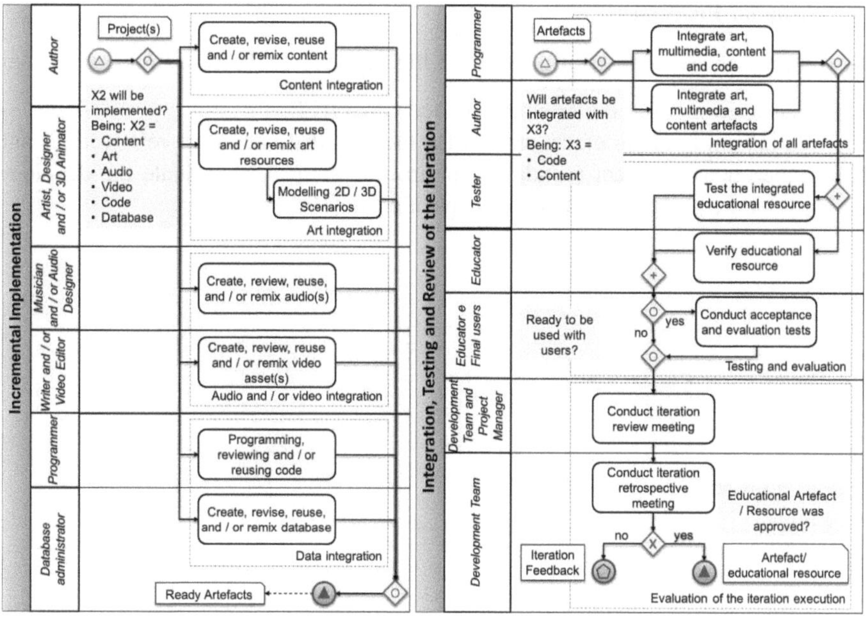

Fig. 4. Overview of the production activities: "analysis and planning of the iteration" and "iterative project" processes.

(8) Integration, Testing and Review of the Iteration: Consists of: (8.1) Integration of all artefacts; (8.2) Testing and evaluation; and (8.3) Evaluation of the iteration execution. Programmer is responsible for integrating the artefacts of art, multimedia, content and programming. Author is responsible for integrating art, multimedia and content artefacts. Tester is responsible for conducting the tests with the integrated educational resource.

3.4 Post-production Processes

In post-production, there are processes performed after that the educational resource was delivered, as illustrated in Fig. 5 [33].

(9) Environment and Maintenance: consists in (9.1) Configuration and installation; and (9.2) Management of change and maintenance requests. It is performed by development team.

(10) Execution: (10.1) Used by the students and educator and (10.2) Measurement and feedback (by the system or educational resource).

(11) Learning Assessment: (11.1) Human performance assessment; (11.2) Self-evaluation and (11.3) Satisfaction rating; that are performed by students.

3.5 Support Processes

These processes are performed throughout the lifecycle of the educational resource:

(12) Verification (were the artefacts created correctly?),
(13) Validation (were artefacts created for the intended use in the real world?) and
(14) Experimental Project (conduct experimental studies to evaluate the developed product, e.g., is it effective?), as specified in [19].

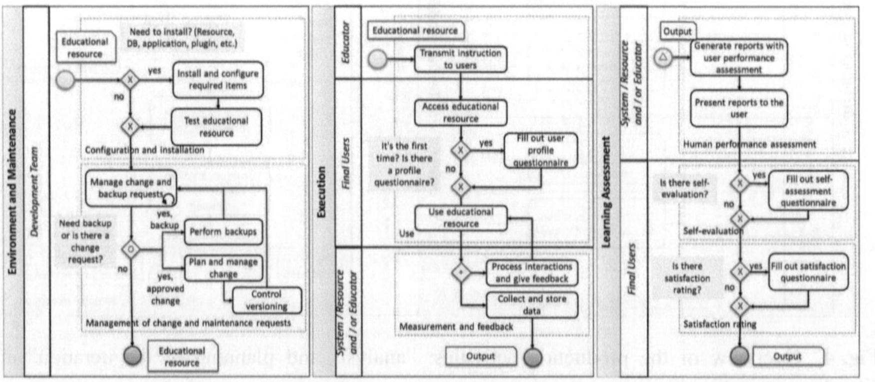

Fig. 5. Overview of the post-production activities: “environment and maintenance”, “execution” and “learning assessment” processes.

4 Development of a Serious Game with AIMED

To verify the feasibility of the AIMED method (evaluate both entire lifecycle and resources produced), it was used to develop a serious game for firefighting, named *FirefightingGame*. Plans, models, specifications and diagrams were created, using the proposed templates, in a collaborative way using Google Drive© tools.

Project Management (Vision): *FirefightingGame* is multiplatform and contains three 2D phases (1, 5 and 6 - with card games) and three 3D phases (2, 3 and 4 - with 3D virtual environment). It was produced and tested by the development team: 01 instructional and game designer, who also played the role of project manager, 01 developer and 01 domain expert (fire captain).

Initial Planning (Initial Plan): (1) Pedagogical requirements: The *FirefightingGame* was designed to train the following competencies: (1a) Knowledge: to know the protocol to fight a fire by indirect attack; (1b) Operational technical skills: to properly use the indirect attack protocol to extinguish the fire, ventilate, rescue victims and leave safely the site; (1c) Attitudes: ensure safe access, extinguish the fire, ventilate the environment and leave it safe. (2) Technical requirements: (2a) the creation of different phases, with different interaction resources (2D game and 3D virtual environment); (2b) multiplatform access with data collection and storage on the server; (2c) the possibility of implementing distributed simulations.

Analysis and Planning of the Iteration: there were 7 iterations (from 3 to 6 weeks each), the 2D phases were created sequentially (1, 5 and 6), after the database and login, and then, the 3D phases (2, 3 and 4). Art resources (images, textures and 3D models) and audio were reused from free repositories available on the Internet and integrated into the virtual environment using Unity3D©.

Iterative Project: the firefighting context and protocol were analysed and projected in the serious game phases, through the integrative models described in the DevJSTA [19]. The integrative model of assessment-training program enabled to include different activities and types of assessment and feedback (Table 1).

Table 1. Integrative model of evaluation-training program (adopted [19]).

	Bloom's taxonomy	Kolb's experiential learning model	Types of assessment	Kirkpatrick' evaluation model	Types of feedback
1	1: knowledge	Concrete experience	Diagnostic	2: learning	Constructive/descriptive
2	1: knowledge	Abstract conceptualization	Formative		Constructive/descriptive/ prescriptive
3	2: understand	Active experimentation			
4	3: apply				
5	4: analyse	Reflective observation	Summative		Constructive/descriptive
6	5: synthetize				
7	6: evaluate	–	Self-assessment	1: reaction	–

The serious game includes seven phases: 1 and 2 should explore level 1 of the Bloom taxonomy, describing objectives that make it possible to remember the trained protocol (concrete experience and abstract conceptualization), as shown in Figs. 6 and 7a. The activities in phases 3 and 4 enable the learner to plan and do actions based on the knowledge that is reflected (active experience) (Figs. 7b and 8a). In them, the learner must be able to understand and apply this knowledge and obtain prescriptive feedback (as he should have done) in addition to the descriptive (correct and wrong). Phases 5 (Fig. 8b) and 6 enable a summative evaluation, and the learner should analyse and synthesize the knowledge practiced from observation and reflection. The last phase (7), the learner should evaluate his/her training and the training program itself, using a questionnaire available on the website.

Incremental Implementation: Unity3D© was used in conjunction with the C# programming language for simulating models and accessing the database. The database was developed in MySQL language and the database server was implemented in PHP. Three card game mechanics were implemented to perform phase 1, and later reused in phases 5 and 6 (Figs. 6a, b and 8b). Behaviours of 3D objects were developed for phase 2 and reused in phases 3 and 4 (such as, open the door and squirt water).

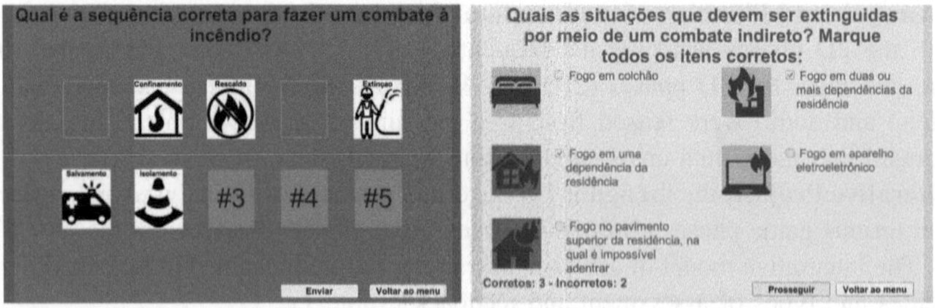

Fig. 6. Serious game screenshots - 2D card game – phase 1: (a) question: "What is the correct sequence for firefighting?" and (b) question feedback: "What situations should be extinguished using indirect attack?".

Fig. 7. *FirefightingGame*: (a) phase 2: 3D residence scenario with fire and steps at the top, in order of execution; (b) phase 3: same scenario with fire, but without instructions.

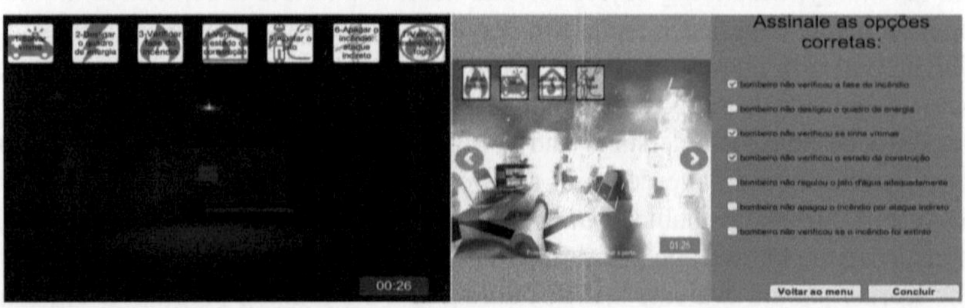

Fig. 8. *FirefightingGame*: (a) phase 4: new scenario with fire and victim; (b): phase 5: scenes for the player to analyse a firefight.

Integration, Testing and Review of the Iteration: the integration of the resources created was done using the Unity3D© game engine (art, audio and programming) and C# (client) and PHP (server) languages.

Licensing: free tools and open source resources from web repositories were used: the game engine Unity3D© (the Personal Licence, free for students), the 3D modelling SketchUp Free© and the 3D Warehouse© repository. In this way, the *FirefightingGame* can be available in an open source version.

Publication and **Environment and Maintenance:** the serious game was developed to run in the web platform, then is not necessary configuration or installation.

Execution and **Learning Assessment:** *FirefightingGame* will be used by firefighters, and then, the self-evaluation and the satisfaction rating will be done using the questionnaire described in [19]. Human performance will be assessed using the data collected by the game.

Verification (developer and project manager) and **Validation** (with the domain specialist)**:** they served as control points at the end of each process, and allowed to improve the phases.

Experimental Project: an experimental study will be conducted to evaluate the effectiveness of the developed product (learning assessment and project evaluation).

5 Discussion

The effectiveness of the educational resource is the result of its production added to its correct use. Thus, the quality of a development method will increase or decrease the effectiveness of the final product. This quality is influenced by processes ("how" must be produced); created artefacts ("what is produced"); project (mainly management); and by the actors (i.e., their competencies to create the artefacts) [18–20].

For interactive educational resources, existing methods are limited by not including all requirements of each area (e.g., game design, simulation modelling, instructional design, software engineering, project management). Moreover, while the method should guide and support the development of the artefacts, in a systematic, standardized and multidisciplinary way, it also should be flexible to cover the different needs of the team members who are using it. In the other hand, methods of the more developed areas (such as, simulation and instructional design) can benefit the development of educational resources. However, if each team member (e.g., instructional designer and game designer) adopts their own method, there may be a lack of integration and/or imbalance of the gameplay and learning effectiveness.

In this context, a method that integrates different areas, and standardizes "what" and "how" should be produced in each step, can help in producing an effective educational resource. This standardization can also enable the reuse of artefacts, and even an automatic generation. However, existing methods emphasize specific requirements of educational resource, or, when more comprehensive, do not address all the requirements for effective learning. These limitations are often caused by the context in which theses educational resources are embedded in other research and development areas, such as, distributed simulation solutions, software engineering, educational games, learning objects, virtual and augmented reality, training and development human performance measurement systems; or in research sub-areas, such as, developing games for specific devices (e.g. consoles or smartphones) or games for particular domains (e.g. health and education).

In this context, the DevJSTA methodology was proposed and evaluated by [19], describing processes, actors' roles, and input and output artefacts; and AIMED extends DevJSTA using agile practices and project management. As reported in Sect. 4, the AIMED method provided support for the iterative and incremental development of the

FirefightingGame. The method artefacts (plans, models, specifications and diagrams) helped in the creation and reuse of the serious game artefacts (learning content, art, audio, database and code). The first phases 3D and 2D demanded more time, and the following less given the reuse.

This serious game (product) was evaluated by domain expert (captain in the Military Firefighters Corps) and the development lifecycle (the use of AIMED) by development team. As future work, an experimental study will be conducted to evaluate the execution, learning (the human performance, self-evaluation and the satisfaction [41]) and project.

6 Conclusion

This project is inserted in the research and practice areas that support the development of iterative educational resources, such as, serious games, training simulations, gamified virtual environment; to understand how they should be planned, developed, evaluated, validated and used for the learning, training and assessment purposes (the entire life cycle).

There is a gap among the areas that support the production of interactive educational resource. Existing methodologies are limited by not including all requirements of each area, in a systematic and multidisciplinary way, and not be flexible to cover the different needs of the team members who are using it.

Thus, in this paper, we propose a new agile method (AIMED), based on the integration of the DevJSTA methodology, with the agile AM-OER method and project management practices.

As future work, AIMED and the serious game developed will be evaluated, as well as, new and different educational resources will be developed and validated. In addition, metrics and indicators will be defined to evaluate the use of the methodology. It should be evaluated qualitatively and quantitatively through case studies with game designers, instructional designers, developers, and domain experts.

Acknowledgment. We thank CNPq and CAPES for the financial support.

References

1. Chung, C.A.: Simulation Modeling Handbook: A Practical Approach. CRC Press, Boca Raton (2004)
2. Maciuszek, D., Weicht, M., Martens, A.: Seamless integration of game and learning using modeling and simulation. In: Proceedings of the 2012 Winter Simulation Conference, pp. 1611–1620 (2012)
3. Messineo, L., et al.: Game-based learning for education and training in the health sector. In: 4th International Conference of Education, Research and Innovation, pp. 3341–3353 (2011)
4. Prensky, M.: Digital Game-Based Learning. Paragon House, Vadnais Heights (2007)
5. Salas, E., Rosen, M.A.: Best practices for performance measurement in military simulation-based training: observations from the field, pp. 1–29. Report, Department of Psychology, Institute for Simulation and Training, University of Central Florida (2007)

6. Rocha, R.V., Zem-Lopes, A.M., Pedro, L.Z., Bittencourt, I.I., Isotani, S.: Methodology for development of serious games: specification of open source tools. Br. J. Comput. Educ. **24** (3), 109–124 (2016)
7. Rocha, R.V., Zem-Lopes, A.M., Pedro, L.Z., Bittencourt, I.I., Isotani, S.: Metodologia de Desenvolvimento de Jogos Sérios: especificação de ferramentas de apoio open source. In: XXVIII Brazilian Symposium on Computers in Education, pp. 489–498, Maceio (2015)
8. Aslan, S., Balci, O.: GAMED: digital educational game development methodology. Simulation **91**(4), 307–319 (2015)
9. Akilli, G.K., Cagiltay, K.: An instructional design/development model for the creation of game-like learning environments: the FIDGE model. In: Pivec, M. (ed.) Affective and Emotional Aspects of Human-Computer Interaction: Game-Based and Innovative Learning, pp. 93–112. IOS Press, Amsterdam (2006)
10. Hays, R.T.: The effectiveness of instructional games: a literature review and discussion. Report, Naval Air Warfare Center Training Systems Division (2005)
11. Zin, N.A.M., Jaafar, A., Yue, W.S.: Digital Game-Based Learning (DGBL) model and development methodology for teaching history. WSEAS Trans. Comput. **8**(2), 322–333 (2009)
12. Arimoto, M.M.: Agile development of open educational resources. Ph.D. thesis, University of Sao Paulo (2016)
13. Becker, K., Parker, J.: Serious instructional design: ID for digital simulations and games. In: Resta, P. (ed.) Proceedings of Society for Information Technology & Teacher Education International Conference, pp. 2480–2485. AACE, Chesapeake (2012)
14. Freitas, S., Jarvis, S.: A framework for developing serious games to meet learner need. In: Interservice/Industry Training, Simulation, and Education Conference, pp. 1–11 (2006)
15. Kirkley, S., Tomblin, S., Kirkley, J.: Instructional design authoring support for the development of serious games and mixed reality training. In: Interservice/Industry Training, Simulation, and Education Conference, Bloomington, pp. 1–11 (2005)
16. Rodrigues, H.F., Machado, L.S., Valença, A.M.: Applying haptic systems in serious games: a game for adult's oral hygiene education. SBC J. Interact. Syst. **5**(1), 16–25 (2014)
17. Van Der Zee, D.J., Holkenborgb, B., Robinson, S.: Conceptual modeling for simulation-based serious gaming. Decis. Support Syst. **54**(1), 33–45 (2012)
18. Rocha, R.V., Campos, M.R., Boukerche, A., Araujo, R.B.: From behavior modeling to communication, 3D presentation and interaction: an M&S life cycle for serious games for training. In: IEEE/ACM 16th International Symposium on Distributed Simulation and Real Time Applications, pp. 132–139 (2012)
19. Rocha, R.V.: Metodologia Iterativa e Modelos Integradores para Desenvolvimento de Jogos Sérios de Treinamento e Avaliação de Desempenho Humano. Ph.D. thesis, Federal University of Sao Carlos (2014). (in Portuguese)
20. Balci, O.: A life cycle for modeling and simulation. Simulation **88**(7), 870–883 (2012)
21. Banks, J., Carson, J.S., Nelson, B.L., Nicol, D.M.: Discrete-Event System Simulation, 3rd edn. Prentice-Hall, Upper Saddle River (2001)
22. IEEE 1516.3-2003: IEEE Recommended Practice for High Level Architecture (HLA): federation development and execution process (FEDEP). IEEE, New York (2003)
23. IEEE 1730-2010: Distributed Simulation Engineering and Execution Process (DSEEP). IEEE, Washington (2010)
24. Ford, F.: The Euclid RTP 11.13 SE development & exploitation process (SEDEP). In: European Simulation Interoperability Workshops, Edinburgh, pp. 1–10 (2004)
25. Chandler, H.M.: The Game Production Handbook, 2nd edn. Bookman, Porto Alegre (2012)
26. Novak, J.: Game Development. Cengage, Sao Paulo (2010)
27. Schuytema, P.: Game Design: Practical Approach. Cengage, Sao Paulo (2008)

28. Branson, R.K., Rayner, G.T., Cox, J.L., Furman, J.P., King, F.J., Hannum, W.H.: Interservice Procedures for Instructional Systems Development (5 vols). Ft. Monroe, U.S. Army Training and Doctrine Command (1975)
29. Dick, W., Carey, L., Carey, J.O.: The Systematic Design of Instruction. Allyn & Bacon, Boston (2004)
30. Kemp, J.: Instructional Design: A Plan for Unit and Course Development. Fearon-Pitman, Belmont (1977)
31. ABNT: NBR ISO 10015:2001 - Gestão da Qualidade: diretrizes para treinamento. ABNT, Rio de Janeiro (2001)
32. Trybus, J.: Game-Based Learning: What it is, Why it Works, and Where It's Going. NMI White Paper. New Media Institute, New York (2010)
33. Rocha, R.V., Valle, P.H.D., Maldonado, J.C., Bittencourt, I.I., Isotani, S.: AIMED: agile, integrative and open method for open educational resources development. In: IEEE 17th International Conference on Advanced Learning Technologies, Timisoara, pp. 163–167 (2017)
34. Rocha, R.V.: Critérios para a construção de jogos sérios. In: XXVIII Brazilian Symposium on Computers in Education, Recife, pp. 947–956 (2017)
35. Rocha, R.V., Bittencourt, I.I., Isotani, S.: Análise, Projeto, Desenvolvimento e Avaliação de Jogos Sérios e Afins: uma revisão de desafios e oportunidades. In: XXVIII Brazilian Symposium on Computers in Education, Maceio, pp. 692–701 (2015)
36. Marklund, B.B.: Games in formal educational settings obstacles for the development and use of learning games. Dissertation (Informatics), University of Skövde (2013)
37. Beck, K., Andres, M.: Extreme Programming Explained: Embrace Change, 2nd edn. Addison-Wesley, Boston (2004)
38. Schwaber, K.: The Scrum development process. In: Proceedings of the 10th Annual ACM Conference on Object Oriented Programming Systems, Languages, and Applications, pp. 117–134. ACM, Austin (1995)
39. Ambler, S.W.: Agile design practices (2001). agilemodeling.com/essays/agileDesign.htm. Accessed Feb 2016
40. Koper, R.: Current research in learning design. Educ. Technol. Soc. **9**(1), 13–22 (2006)
41. Rocha, R.V., Bittencourt, I.I., Isotani, S.: Avaliação de Jogos Sérios: questionário para autoavaliação e avaliação da reação do aprendiz. In: XIV Brazilian Symposium on Games and Digital Entertainment, Teresina, pp. 1–10 (2015)

Author Index

MIX
Papier aus verantwortungsvollen Quellen
Paper from responsible sources
FSC® C105338
FSC
www.fsc.org

If you have any concerns about our products,
you can contact us on
ProductSafety@springernature.com

In case Publisher is established outside the EU,
the EU authorized representative is:
Springer Nature Customer Service Center GmbH
Europaplatz 3, 69115 Heidelberg, Germany

Printed by Libri Plureos GmbH
in Hamburg, Germany